The Chronosphere Commentary

Alan AtKisson

BROKEN BONE PRESS

Copyright © 2020 Alan AtKisson

First Edition published by Broken Bone Press

All rights reserved.

Cover image:

Detail from "The Curve", acrylic on canvas, by Alan AtKisson

ISBN: 978-0-9911022-6-6

DEDICATION

To Robert K. Meadows,
who helped me see
who I might become.

Introduction

In the fall of 1997, my friend Wouter Biesiot was dying. He had been battling colon cancer for several years. So I began to write.

Wouter was a nuclear physicist by training, but he had switched career paths and begun working on sustainability, inspired partly by the work of our mutual friend Donella "Dana" Meadows, and her former husband Dennis Meadows. Wouter (pronounced "VOW-ter") and I used to meet annually, in Hungary, at meetings of the Balaton Group, a network of systems thinkers and sustainability practitioners started by Dana and Dennis. We became friends, and we had long discussions on questions of philosophy and the purpose of life, in addition to sustainability topics such as lifecycle analysis, a topic on which Wouter was a technical pioneer.

Wouter had credited me with helping him decide to spend a good deal of his remaining energy, during the last half-year of his life, composing a book called *Fragments of a Dream*. In actuality, all I did was encourage Wouter when he confessed to me a strong need to write such a book. The book summarized (I am told, for I could not read Dutch) Wouter's technical ideas, as well as his personal thoughts about the sustainability challenges facing the Netherlands. *Fragments of a Dream* was Wouter's final contribution to the world of sustainability that he cared so much about, a last mark in the sand before he himself was swept away into the ocean of time.

During the period of his illness, I visited Wouter's home in Gröningen, in the northern part of the Netherlands, three times. The first time, Wouter was in between rounds of chemotherapy. We rode bikes together, drank a beer in town, debated cosmology with one of his physicist friends, and went to the sea with his partner, Nanda. On my second visit, Wouter was a frail and skinny rail of a man, but he was able to walk slowly around the art museum, if he took frequent rest breaks.

The third and final time I visited Wouter, he was on his deathbed and receiving regular doses of morphine from Nanda, to control his pain.

It was on this third and final occasion of visiting Wouter in his home that I presented him with a hand-written copy of *Chronosphere*, the poem that his battle with cancer had inspired. I had barely finished in time to give it to him. He could barely hold the small chapbook, so I read it to him, and we talked a bit about the meaning of the poem, which has many references to experiences and ideas we had shared.

About two weeks later, Wouter passed away.

Many years later, one of his daughters (who had been a child at the time of his death) wrote to ask me about *Chronosphere*. She found its meaning opaque. I wrote back to explain some of the basic ideas, contexts, and references — and that gave birth to this commentary.

And now, dear reader, I will try to explain *Chronosphere* to you.

*

Chronosphere circles around the human relationship to time. As of 2018, the term "chronosphere" had entered popular culture through a Disney film version of Lewis Carroll's *Through the Looking Glass:* Alice steals a chronosphere device from the Lord of Time, at the heart of a huge contraption that controls time's movements — but this was an invention of the movie-makers and not present in the original *Alice in Wonderland* books. If you do an internet search, you will find that there is also a thrash-metal rock band, a card game, and a complex data management service all sharing the name "Chronosphere".

But when I wrote this poem, in 1997-1998, I believed the term to be my own invention.

During the period when Wouter's life was winding down, I was quite obsessed with the topic of time, partly because I had responsibility for organizing a five-day scientific meeting on that theme. Time, at least in all the normal ways that we subjectively experience it and conceptualize it, is not simply a feature of the universe in which we live. It is a collection of widely varying cultural conventions, perceptions, beliefs, psychological experiences, and of course measurement systems that scholars from many disciplines have studied for, well, a long time. I began to visualize time as a sphere comprised of all humanity's clocks, calendars, time-related technologies, and thoughts about the subject, encircling the planet, something like the atmosphere, the biosphere, or Vernodsky's noosphere. We live in a bubble of time, just as we live in a bubble of air, life, and consciousness — all packed into the fragile outer membrane of a small, rocky planet.

But time is also an undeniably objective phenomenon that carries us relentlessly forward along its arrow, from past to future, birth to death. The structure of *Chronosphere* attempts to sketch humanity's complex relationship to time and to reflect two of its most crucial, structural features, as experienced by human beings: time circular and cyclical, on the one hand, and linear and one-directional, on the other.

Thus the poem is arranged cyclically, in 61 stanzas, where number 61 echoes number 1 (think 60 seconds or minutes) and suggests an endlessly repeating clock-like pattern, with small variations. The seasons repeat themselves, for example, with great similarity and regularity … but with differences. Winter is winter, but each new winter varies a bit from the previous one.

Chronosphere also has a linear direction. For the Falling Man — the main protagonist, inspired by my friend Wouter, whose time was running out — that direction is down. His counterpart, the Flying Woman, does not share the same fate: she moves around. Her indifference to the Falling Man's terminal fate is like the perceived indifference of time itself to the travails of humanity. Her aggressiveness says something about how we (or at least I) experience that indifference.

These two character-driven narratives interweave, but do not fill the whole space created by Chronosphere, which includes poetic forms ranging from haiku to free verse to rhymed quatrains. Some verses drive the character narratives forward. Others play with aspects of how humans relate to time, the words we use, the many concepts we have created to break up time's monolithic monotony into prismatic variety: "moment," "hiatus," "millennium", "holiday".

Wouter would mostly not have needed any explanation of the references to system dynamics and other scientific or academic concepts that were commonly discussed in our Balaton Group Meetings at that time. He would have had his own experiences of both the physical Lake Balaton and the group dynamic to draw on. But since *Chronosphere* is made opaque to some readers by these references, I will unpack many of them in the commentaries that follow.

Chronosphere can either be seen as one long poem, or many short ones strung together. The verses are like moments, discrete, but connected — not unlike the way we experience our lives. So we will take a tour through these verses now, one by one.

Let's begin. Or not. The choice to keep reading is yours.

1.

Lake Balaton

is a sheet of blue glass,

a window between after

and before. I swim

through the pane. On shore,

we hear the sound of laughter.

The future looks clear; the past

remains opaque. Down

at the bottom of the lake,

I see graffiti in the sand:

"The Chronosphere

is in your hands."

Commentary

"Pane" is a terrible pun — a *double-entendre* with a dual purpose.

The pain that Wouter was experiencing was terrible. I remember him saying to me: "I tell myself that the pain is really only happening right now. It was not happening before, and it certainly will not be happening long into the future [because he knew he was going to die within a couple of months]. It is really just a brief moment, this pain."

(Did Wouter actually say this? Or did I manufacture the memory, based on his stoic attitude?)

This first verse differs from all the other verses in Chronosphere. It introduces an "I" that does not return (in this form) until verse 61. It launches the poem, a separate narrative that fills all the other verses; but it is somehow set apart from it (as is verse 61). Verse 1 is self-referential: not only is the Chronosphere — the sum total of all human time, as lived up to this moment of your life — in your hands; so is this piece of writing about it.

Why Lake Balaton? This poem contains numerous references to a scientific meeting that happens annually on the shore of a large Hungarian lake. The people and topics discussed at the "Balaton Group Meeting" are important, but there is also the physical lake itself: often very still, its great flat surface demarcating a boundary between two worlds. "After" and "before" are like two universes, to which we humans never really have access — no matter what the physicists say about time's essential integration with space.

But the physicist's tale of time is not ignored. A lake, unlike a river, gives us the image of a "pool" of time, something static and space-like. Let's say the future is above the surface: this is where the sun shines, this is where most of our thoughts aim to be. Our imaginings about the future are much clearer, oddly, than our memories will ever be.

Certainly the future was clear for Wouter. He, like all of us, could dive into memory, rummage around in it, find a few things if they were needed (and stumble upon others). But in general, the past is surprisingly difficult to access, and constantly recedes, whereas the future is rushing towards us, getting bigger — and more certain — all the time.

What else should one say about this verse? Students of poetry will notice internal rhymes (before, shore; lake, opaque) and alliterations (the r's in future and Chronosphere, the f's in future, graffiti, laughter, and of course sphere). But these efforts to imbue the text with some musicality of language, some attention to rhythm and form, are to be expected as part of the effort to construct a serviceable poem out of a jumble of thoughts and images linked to a scientific meeting and the dying of a friend. These poetic commonplaces will not otherwise be mentioned in the rest of this Commentary, unless there is some aspect of the language or poetic form that is integral to the meaning and non-obvious to the non-poetically inclined reader.

2.

What is time?

Dynamic.

(s)

Commentary

My understanding of time, at this juncture in my life, was intensely affected by the systems modelers in the Balaton Group, for whom dynamics are central. Time *is* dynamics: the process of unfolding change, driven relentlessly forward by an ungodly number of causal reactions and their spin-off side-effects and feedback loops. This is the complexity of the world that friends like Wouter tried, improbably, to model in equations.

But that word, dynamic, has many other associations that give it an explosive punch. The Batman and Robin of my childhood, for example, were called "the Dynamic Duo". A creative or influential individual can be described as "dynamic."

Time is *dynamic*: it hits us with motion, energy, change.

Time is *dynamics:* the sum total of all those changes.

The most important line in this verse is the "(s)", with its signal of ambiguity and ambivalence (it makes the word in the previous line swing either way) and its slight resemblance to a term in an equation. The "(s)" is an in-joke, a play-on-words that systems modelers are likely to understand quickly. Almost anyone else is likely to need this explanation.

When the verse is read aloud, one asks the question "What is time?" and declares the answer boldly and confidently: "Dynamic." But then, after a beat, comes that little hissing, *sotto voce* "(s)". Suddenly, the certainty and simplicity of that answer is undermined, just as real life — say, a diagnosis of terminal cancer — undermines the simplicity of time's treatment in the equations of a model.

Time is dynamic, and dynamics. Both. And much more besides.

3.

Many futures, or one?

One past, or many?

Too few futures.

One too many pasts.

Commentary

Philosophical debates on the nature of time, throughout the ages and presently, have often honed in on the question of destiny. Is our future predetermined? Does it simply sit there, as unchangeable as a rock in the river, waiting for us to reach it? Or are there multiple branching options? Do we have the freedom to set our own course and create our reality? Or are we cogs in a deterministic machine?

The debate, originally religious, has lately been dominated (in the popular, educated Western mind) by physics. Some present the mind and its choices as the result of (ultimately predictable) physical processes of cause and effect, thus making choice, and even consciousness itself, an illusion. Others speak of multiple universes, each coming into reality by the branching pathways of decisions made in a sea of probability. In one universe, you chose to eat that second helping of lasagna, while in another — the one you are experiencing right now as "real" — you did not. The "you" that ate the lasagna also experiences that universe as real. A simplified caricature, but this is how some physicists believe reality works.

The past, meanwhile, once looked solid and monolithic. It is the path we already walked, the river we already navigated. How could there be more than one of these? Ah, but let us begin comparing notes, and soon we will discover that our memories and experiences of the path differ. There are as many versions of the "past" as there are perceivers, with their widely differing values systems and filters, to perceive it and describe it. Even systems modelers, at the time this poem was being composed, were experimenting with building multiple versions of the past into their equations, based on the most common groupings of people in society and their differing interpretations of "what really happened."

The debate on whether our future is determined or up for grabs has kept neuroscientists and physicists and philosophers squabbling for decades. But the answer to the two questions posed in the first half of this stanza is, for the sake of this poem, once again inflected by the experience of my dying friend Wouter. Regardless of what the physicists or the neuroscientists might say, the variety of Wouter's potential futures was severely constrained; all of his pathways led to death. And the multiplicity of his pasts was spoiled by that unavoidable moment when he was diagnosed with cancer: this, his real past, was "one too many," a phrase tinged with regret.

Though cancer chooses us, it is hard to escape the feeling, Wouter once told me, that we have somehow chosen it.

I do not want to overstress these links to Wouter's life and death, because the poem is not just about that. While *Chronosphere* attempts to reflect and honor the extraordinary transparency and grace with which one person, my dear friend, described his experience with a terminal illness, it is also reflecting on the trajectory of humanity — and indeed the Earth itself — through time. At this moment, with climate change accelerating and resource problems mounting, our global futures may also be constrained. We may be contending with a set of avoidable choices that called into being this timeline. We might, as a civilization, have chosen otherwise; but now we have "one too many pasts."

4.

A man's life teeters

on the brink. He has

no respite time

at all. His response

is to fall.

Commentary

During a presentation at the annual Balaton Group Meeting on time and sustainability, Wouter Biesiot (his last name is pronounced "Bee-zee-OH") introduced us to two concepts borrowed from his previous work on nuclear power plants. "Respite time" is the time you have left once an accident has started to occur: a crack in the cooling system, an operator error. Such moments start the metaphorical ticking of a timer, and unless one acts in time — that is, within the window known as "respite time" — a disaster will occur.

"Response time" is the time it takes you to initiate and complete an adequate response, one that rectifies the situation and avoids, say, a meltdown and/or radiation leak.

Respite time and response time are enormously useful concepts for thinking about any situation where choices or accidents can cause catastrophic problems, including global sustainability problems. Take ozone depletion: the invention, large-scale production, and release into the atmosphere of chlorofluorocarbons (CFCs) started a process of destruction in Earth's protective ozone layer. By the time this process was discovered by Nobel Prize-winning researchers Sherwood Rowland and Albert Molina, there was a limited window of time available — the respite time — in which to act before the process reached a point of no return. Left unaddressed, this problem could have ultimately weakened or even undermined human civilization on a global scale, because of the resulting high levels of ultraviolet radiation that would stream down to earth and impact both health and food production.

Fortunately, in the case of CFCs and the ozone layer, humanity's response time — the time that it took us to mobilize, change, and avert disaster — was *shorter* than the available respite time. Representatives of government and industry made decisions, created policies, and transformed technologies. We changed *in time*.

But in other cases — the release of greenhouse gases like CO_2 being a prime example — our response time has been *longer* than the respite time. We are not acting quickly enough. Unless we accelerate, disaster is inevitable.

(During the 2011 Fukushima nuclear disaster, the whole world witnessed what happens when catastrophe suddenly shrinks respite time to nearly zero. The fastest response time in the world could not have prevented that triple meltdown.)

For a person diagnosed with terminal cancer, there is no respite time. There is no action he can take to avoid the predicted disaster. Indeed, there is no "respite" available to him either, in the other, more common sense of that word: a break, a reprieve. Wouter, who fought off cancer as long as he could with every technique worth trying, from chemotherapy to the prayers of friends, understood and accepted that his fate was unavoidable.

Thus he began falling, gracefully, towards his own death.

5.

The Falling Man passed

a Flying Woman. He cried,

"Please toss me a rope!"

"Get lost," she replied.

Commentary

This verse, like most verses in *Chronosphere*, can be read and interpreted in several different ways. It is not advisable to try to explain them all. But the characters presented here will be with us for the rest of the poem, and some introductions are in order.

The Falling Man of *Chronosphere* is not Wouter; he is an "everyman," *inspired* by Wouter.

Why "falling"? A fall is an irreversible, one-directional process. It is governed by the laws of physics; so are our lives. All of us are ultimately falling towards death.

But while falling, we can still be who we are. Consider Alice in Wonderland, falling down the long, long rabbit hole: she notices that she still "has time", and she uses it. She looks at the things on the walls of the rabbit hole as she passes by them. She thinks to herself. She recites text memorized from a book. She wonders what will happen to her later. Falling is just the state she is in. The end of the fall may be unavoidable, but she has considerable freedom to be herself, up until that sudden moment.

And yet ... and yet ... we do not want to die. We do not want to reach the end of the fall. And we somehow notice, all around us, processes that do not seem to share our brief, one-way adventure. We may know theoretically that one day the sun will expand, consume the Earth, and ultimately burn out. But that is billions of years into the future. The solar system's fall is much, much longer than our own — so long that it might as well be an eternity.

Meanwhile, we see evolution continuing. We see the cycles of nature, the continual birth and rebirth of all other living things. We see technological advance stretching across the generations. We see those other kinds of time, to which we can relate, and occasionally even identify, and we reach out to them, as though reaching out for help. Our minds take flight to grapple with these other ways of knowing and experiencing time. But these non-human forms of time are ineluctably impersonal, distant from us, unreachable. Ultimately, time itself does not care about our little, short-lived fates. It swoops past us, mercilessly.

Time, we say, flies.

That is the best I can do by way of explaining the Flying Woman.

6.

The Flying Woman

wrote a letter

to the Falling Man.

"That's all," she wrote.

Commentary

Oh, how very much can be read into the short phrase, "that's all." But to explain every verse in Chronosphere in such detail would be tiresome in the extreme. Time, after all, does not say very much. It is not mute; just perfunctory.

And sometimes, it appears to be making bad jokes.*

Chronosphere is a playful poem: it plays with words, with forms, and at the time it was written, it was also playing with a new technology called the World Wide Web. The first version of *Chronosphere* was conceived for publication on the web, with hyperlinks in every verse that took the reader off to some interesting, time-related website.

This quickly proved to be a nightmare for me, as the poet and website manager, because even in those heady early days of the Internet, websites came and went like flowers, blooming and wilting. Links went dead, sometimes within just a few weeks of making them.

What verse 6 once linked to, I do not recall, but it was a letter of some kind. What is left is an expression of the Flying Woman's capricious character: she is by turns approaching and rejecting the Falling Man.

Awakening hope … and then dashing it.

* The phrase "That's all she wrote" means "It's all over" and is thought to have originated during World War II, when many American soldiers received break-up letters from their girlfriends back home. These letters came to be known as "Dear John" letters. After a while, the phrase "Dear John" was enough; it meant the relationship was finished. An old joke I once read is that one soldier received such a letter and was encouraged to read it aloud to his friends. "Dear John," he read. Then he stopped. "Well, go on!" they said. "There isn't any more," he replied. "That's all she wrote."

7.

The Falling Man

looked down. He was definitely going

to hit the ground, but it seemed

so far away. How long?

It was impossible to say.

Commentary

No matter how close we are to death, it always seems more distant than it actually is. I cannot say that I know this from personal experience — at time of writing, I have no idea when I might die. But I have been close enough to death, to the fact of someone dying, enough times in my life that I believe this to be so. Hope holds the reality of death at bay (see verse 8).

I do not just mean that even in old age and illness, a nearing death feels farther away than it actually is, even when we know it is coming. I also mean that we have no idea how close death is, at any time, and that we routinely assume that it is quite distant. For example, the co-founder of the Balaton Group, Dana Meadows, died suddenly and unexpectedly in 2001. She was working on a newspaper article. She experienced a headache, which she ignored. Two days later, she was gone, struck down by meningitis.

Meanwhile, as long as we are alive, we feel alive. The Falling Man does not feel any less alive for being closer, at every second, to his inevitable contact with termination. This combination of life and death's binary nature (we are either alive, or not) and the uncertainty and the assumed distance, in our imagination, of death's arrival point is a great gift to us. For to imagine the opposite — to really feel one's life diminishing and death approaching, in anything other than an imaginative or poetic way — is often experienced as a nightmare.

Think of it this way: the reality of our impending death is managed far less cruelly by time than it could be.

8.

The Falling Man fell

asleep. Sleep made love

to him all night long.

He awoke

pregnant with hope.

Commentary

What happens to us when we sleep? Normal consciousness is suspended. We are still, sometimes, aware of ourselves. But consciousness is almost always consciousness *about* something. We may become aware that we are dreaming. Or we may become aware that we are lying in a bed, and that we are "on the verge" of sleep.

Otherwise, in sleep, it is as though we cease to exist. That is why sleep has often been called the "little death." So has orgasm: in French, it is *la petite morte*.

Sleep, then, can be a great comfort, even for those who are dying. While asleep, even death ceases to exist. And when we wake up again, it is always a return to life, as though something in us has been reborn.

9.

Meanwhile:

the boys went to the park

and kicked around

an old hiatus they'd found

in the attic, packed away

among the battered moments.

Commentary

Here the poem takes its first diversion, away from the Falling Man narrative. "Meanwhile" means while something else is happening. It hints at simultaneity: events occurring at the same time (more or less), but in another place. Even when one is dying — a process that seems to collapse the whole of the universe into me, here, now — there is always something else happening, elsewhere. A great many somethings.

Here I was also starting to play, explicitly, with words about time. "Hiatus" is an old word for a pause, a break in the continuity in a sequence of actions or happenings. "There was brief hiatus in the war with France," says the Google dictionary, by way of example, which simultaneously (but not meanwhile) gives us a sense — because of the historical overtones in the phrase "the war with France" — that "hiatus" is a word belonging to another time. And this may be true, for "hiatus" is Latin in origin, and its use has been steadily declining since the 1970s.

With no hiatus in sight.

10.

Time's a liar,

Time is true,

Time has nothing

much to do.

Commentary

The concepts and measurements of time with which we surround ourselves are not, in any fundamental sense, real. We made them up. And yet we must abide by them, if we are to live in harmony and cooperation with others.

This word "abide": we use it to mean to mean follow, as in "abide by the rules". But its original meaning is to wait. To stay somewhere. Sometimes it means, to *live* somewhere: one abides in an abode.

Einstein's theory of relativity conjoined space and time into spacetime: a four-dimensional, geometrically homogenous reality, a "place" if you like, where "length of time" is no different than what we think of as physical distance. We may experience time as *moving*, but in fact, it is not doing anything. We are.

We must live with our concepts of time — or perhaps *in* them — whether they are true, or not. To find out, we must wait.

11.

At the airport of time,

developments were taking off.

Changes were landing. Now

and then, they launched

a true evolutionary leap.

Commentary

If there is one place in our globalized and industrialized civilization where time seems to have come to a standstill, it is the airport.

Airports, despite being governed by flight schedules, seem to have paused at a moment of continuous-yet-similar motion: people walk by, radiating a sense of purpose, pulling rolling suitcases. Some sit reading. Others are eating, drinking, talking, waiting. There is always someone glancing around, looking lost, and someone else who appears to be in a hurry. As in an anthill, people mill about, there is constant motion; and yet nothing in the pattern appears to change.

In the midst of this stillness-in-motion, the verse puts nouns about change into the role of airplanes (and perhaps one is a rocket, since it is "launching"). We generally use words like "development" to describe unfolding processes; but here, they are objects, things in motion. Things that fly. Accelerating, slowing down. You can watch them, going and coming.

The phrase "now and then" calls attention to itself here, across a line break. In literal terms, it simply names two points in time: the present, and some other undetermined past or future moment. By naming just these two points, we actually mean something quite remarkable, something that is reflective of our unique mental ability to consider many things at once, and their relations to each other. We are saying, "I perceive a pattern, a frequency of occurrence, discernible in the landscape of time."

Airports are wonderful places to think about time.

12.

The Flying Woman sensed a shadow

behind the deadlines, some great evil

hidden

inside the schedule, but

she exorcised it

with a few strategic delays.

Commentary

Surely all of us have felt, at one time or another, the tyranny of time-planning tools. Here the Flying Woman is given some agency. It seems she has her own story, distinct from the Falling Man's. She observes. She intervenes to keep order. She sets things right by making them slightly wrong. A "delay" is a negative thing, from a time-planner's perspective. And yet delays sometimes work to the advantage of nearly everyone involved in a project or event.

Here we might begin to wonder: can the Flying Woman save the Falling Man, after all?

13.

When the Flying Woman dashed

across dynamic traffic, the cycles

screeched to a halt.

Commentary

Not only does the Flying Woman have agency (verse 12); her appearance alone has impact. She can stop whatever is happening in its tracks.

Now, let's leave aside the lyricism in this stanza (even though the music and rhythm of these lines might be the main point). Pass over the puns. Remember, the first and principal audience of this poem was a community of systems modelers. The Flying Woman has already proven that she can disrupt their world. She can also disrupt their models.

Chronosphere is not just about time; it is about our *models of time*. We see change (dynamics). We see recurring patterns (cycles).

The Flying Woman will return many times to remind us that what we see, when it comes to time, is not equal to what is.

Nota bene: Many of the most successful system dynamics modelling exercises that I was acquainted with in the 1990s had something to do with analyzing and trying to improve urban traffic patterns, often by increasing accessibility for bicycles.

14.

The Flying Woman had been hunting

through the cupboards for hours,

found a few stray pauses, a couple

of interludes. No use;

there was nothing but

historical residue.

Commentary

What is going on with the Flying Woman? What is her agenda? She stiff-armed the Falling Man several verses ago, with great indifference. Now she seems purposeful, on a mission.

"… for hours": Does this mean she was searching for a long period… or that she was she looking for objects, called "hours", that may have been stored away in cupboards?

Perhaps she is searching, specifically, for *longer* periods of time, but finding only short ones — and left-over short ones, no less, old remainders. Mere breaks in the action, shards from an earlier, used-up time ("history"). She finds no larger bits of fresh time, something she can actually use.

What does she want to use it for?

15.

"We turned off the actual world,

and tried running the simulation

for a thousand years.

Most of that time,

life was pretty damn good."

Commentary

This is a direct quote (more or less) from an actual presentation at the Balaton Group Meeting, sometime in the 1990s. What the modeler meant was, they had been verifying their simulated world — a fictional place where they could test some sort of policy or planning intervention — against the historical data and dynamics of the real world. Modelers generally do this. If the simulated world produces dynamics that parallel the real data from the real world, the modelers have more confidence in the fact that their simulations are useful. The experiments they run, in the simulation, might actually say something meaningful about what would happen if the same intervention was tried in the real world.

The modeler spoke this line with a mixture of satisfaction and irony. What a pity, she or he seemed to be saying, that we could not replace the real world with the simulation — where life was demonstrated, or at least predicted with confidence, to be "good" for a whole millennium.

Our computer models, including today's graphically rich video games and virtual environments, are like parallel worlds with their own timelines. The test simulation mentioned in this verse could play a thousand years in a few seconds. In many virtual reality games, time can be paused, sped up, slowed down.

My children have spent a great deal of time playing a computer game called "The Sims". They create virtual "people", who create families, build houses, have children, work, retire, die. An entire generation can live out its life in what seems, for us, a week or so.

And most of the time, life is pretty damn good.

16.

The tourists were inspired

by the millennium's vaulting

architecture, its stained

past. They sat on

the ancient benchmarks

and admired the milestones.

Commentary

Often, when visiting a place, we are also visiting another time. A tour of the great cities of Europe is as much an encounter with history as it is with culture and geography. To walk through a cathedral is to walk through a thousand years.

Yet one can also see, enshrined within the brickwork and masonry, an older version of the future: for there was a time when the raising of each historic structure was merely an intention, which then became a plan. Each moment of achievement — each benchmark, each milestone — is preserved. That we admire it today is proof that yesterday's vision was realized.

But not without a certain amount of heartache, loss, and disillusionment, which also accumulates over time. These "stain" the past, but they also, in hindsight, add a horrible sense of beauty.

17.

The Flying Woman's friends flocked

to her cloud. "Come dance until

the summer ends!" she trilled.

Just then, the clock struck equinox.

Commentary

Normal household clocks do not "strike" equinox or solstice. But some types of clocks do: Stonehenge, for example, could be seen as a large clock for marking the seasons. In fact, we are surrounded by "clocks" that mark other types of time than the 24-hour day. The calendars on our smartphones are clocks of another kind, plinging in our pockets to remind us of appointments, holidays, shopping lists. Nuclear clocks on satellites communicate with Earth-based computers; one can almost hear the digital equivalent of chiming in their multi-decimal, nano-second accuracy. Telescopes, when they gaze into galaxies, are peering back into time, essentially turning all the light of the universe into a giant clock that measures time in units of a billion years.

More and more clocks, of all kinds, measuring all scales of time: it seems the density of timekeeping on our spinning planet is growing faster than the human population. The Chronosphere is expanding.

Meanwhile, is the Flying Woman's dance just ending, or just beginning? We know only that the dance will conclude when the summer ends. We do not know if it is the Spring or the Autumn equinox that has struck — nor even in which hemisphere, north or south, her cloud is to be found (which also makes a binary difference).

Sometimes an event in life feels like it has just begun when suddenly, it ends.

18.

"So Big Ben bellies up

to the great bar of time.

'I'll have a draft

of your best laughter,'

he intones, soberly.

The timekeeper turns

him down. 'You missed

last call.' The huge bell

throws his weight around.

A fight ensues. Big Ben

loses. The timekeeper

knocks his clock off."

Commentary

This verse has the feel and form of a tale told at a local bar, about something that happened another time, at another bar. Big Ben is, of course, the most famous clock tower in the world. Let's assume, presumptuously, that it is also a symbol for a bygone empire. Even empires run out of time. If they resist, they are rudely reminded of one of the implacable laws of history: when your era is over, it's over.

19.

Dawn sun strikes Angkor.

Thoughts of time arise and fall

like dust motes at war.

Commentary

This verse is haiku.

But I am willing to bet

you already knew.

20.

Time is somber,

Time is gay,

Time has nothing

much to say.

Commentary

We are now one-third of the way through (or around) the Chronosphere. This is the second of the little rhymes that appears every ten verses. "Time has nothing," the text reminds us. There may be "much to say" about that, but time's not talking. We are left to wonder about time's silence, whether it is bearing witness to (or just bearing) something horrible, or happy.

21.

The Cenators were sitting

in council. The situation

was grave. They achieved

consensus in just one round:

the Twentieth Century

could not be saved

Commentary

Composed in at the end of the 1990s, *Chronosphere* is a *fin de siècle* product. The century, you might say, was dying — a situation underscored by the ham-fisted humor (it permeates the entirety of the text) that peeks up here in the form of a *petit jeu-de-mot*: "grave."

"Cenators" are, of course, senators of a time rather than a nation ("century" plus "senator"). But it was nations that doomed the 20th century to infamy: the list of depredations committed by national governments is long, grievous, impossible to redeem. While the century will be remembered for technological breakthroughs and advances in the struggle against poverty and scarcity, it will also — and perhaps mostly — be remembered for the brutality unleashed by those technologies, in the name of state-sponsored ideology and imperial interest.

These are the facts of history, and when considered objectively by an august body such as the "Cenators", the judgment of history must condemn them.

22.

The Buddhists say,

death is certain,

time of death

is uncertain.

The Dynamic Modelers say,

systemic collapse is certain,

time of collapse

is uncertain.

Commentary

The "Dynamic Modelers" referred to here are practitioners of the science of system dynamics (see the commentary for verse 2) who build computer models. These models are mathematical simulations of complex pieces of the world and how those pieces change over time — how global fish stocks respond to intensified fishing, for example, or how the interactions between farming and the financial system might ultimately affect forest cover. These simulations, which attempt to give insight into what might (theoretically) happen in the future under certain conditions, can be large or small, global or regional in their scope. And in 1972, for the first time, a group of system dynamics modelers attempted to simulate the whole world.

The result was the runaway bestseller *The Limits to Growth*, which sold millions of copies and sparked a fierce global debate on humanity's economic and environmental future that is still raging to this day. (I tell the story of this book and its impact in my first book, *Believing Cassandra*.)

In a simulated model of a piece of the world, when things go well, systems attain (or sustain) a kind of well-tuned balancing act among all their elements that is called "dynamic equilibrium." They don't "stay the same" — they are still continuously changing, just as the world continuously changes — but they change *within limits*. For example, the fishers may fish a bit more or a bit less in a given year, the fish stocks may go down or up a bit in response (after a delay), but over time, the level of fishers and fish remains stable. This is basically true of the real world as well, except that the real world has many more elements, making it far less predictable.

But if you push one of those simulated system elements too far — send too many fishing boats into the sea for too long, dump too much waste into fish breeding areas, etc. — the dynamic balancing act tumbles off its tightrope. Fish stocks plummet, fishers lose jobs, people go hungry.

That simulated outcome is called "collapse." Of course, this has been known to happen in the real world as well. That's the point of simulation modelling: to try to avoid unwanted outcomes in reality.

In 1972, *The Limits to Growth* attempted to alert decision-makers to the fact that the authors' simulation models of the world were unable to sustain that delicate balancing act over the long term, unless certain fundamental changes were made in the underlying assumptions. Their simulated world always collapsed, eventually, driven by growing populations, consuming ever more resources, creating more and more pollution, on a limited planet.

This verse is making an ironic equivalence between collapse and death. It is also drawing a parallel between modelers and Buddhist philosophers. Neither comparison is fair or accurate. Modelers are scientists, not philosophers, at least when they are reporting on their models. "Collapse" can be catastrophic for those who experience it — ask any fisher whose fish have disappeared — but it leads to reorganization of the system's elements, and some new system pattern. Death, on the other hand, leads nowhere.

But poetry is not about fairness and accuracy. I wrote this stanza after listening to modelers at an academic meeting. They may themselves have drawn the comparison between the behavior of their model of the world — things always collapsed, no matter how optimistically they treated technological advancement etc., but they could never predict exactly when — and the Buddhist comment on death. Academics have been known to interject ironic comments of this kind into their presentations from time to time.

My friend Wouter, facing actual death, might have appreciated the irony a bit more deeply than most.

23.

After the collapse, they asked

for our advice.

We laughed.

Commentary

Chronosphere continues to indulge in irony, which is the only credible source of humor available to those who are suffering or are facing death.

It is also the humor-of-choice found among scientists whose models foretell grave consequences that are subsequently realized. In the 1970s, 80s and 90s, the universal experience of global modelers trying to warn decision-makers about the risks of humanity's current trajectory was that the modelers were ignored. Or mocked. Or attacked.

Not until the mid-2000s, and the breakthrough in public acceptance of difficult facts that ultimately won both Al Gore ("An Inconvenient Truth") and the Inter-governmental Panel on Climate Change ("IPCC") a Nobel Prize, did the world begin to seek the advice of these scientists. Sometimes. On some topics. And often it still prefers to ignore that advice. Even after reaching international agreements, based on that advice, that are intended to ward off doom.

Usually, however, warnings about impending doom are simply ignored until doom actually arrives. Unless it doesn't. Which generally means someone has listened to the warnings and taken effective action, thus avoiding doom.

Ironically, the successful avoidance of doom makes the "doomsayers" wrong. But that was their intention: most doomsayers *want* to be wrong. Or to be made wrong, thanks to doom-avoiding action. (This complex and paradoxical situation is called "Cassandra's Dilemma." I wrote half a book about it.)

So, and finally, why the ironic laughter in Stanza 23? Because when doom does arrive, asking the question "How could we have avoided doom?" strikes one as rather funny.

24.

Two huge sumo wrestlers

speed across Austria

seeking an economy

of scales.

Commentary

Professor Hermann Knoflacher at the University of Vienna is best known as a promoter of pedestrianization — making cities more walkable. He is also a distinctly charming man, with a glint in his eye, a sharp but self-deprecating wit, and a pronounced disapproval of the apotheosis of the automobile and what it has done to warp our sense of perspective. I recall many of his lectures, delivered in a clipped Austrian accent, with pleasure: they were always eye-opening, and introduced me to ideas that later became commonplace in the world of transportation studies. For example, he showed photographs of people walking around on city streets bearing complicated contraptions built of balsa wood and string, hanging from straps on their shoulders, that represented the amount of space taken up by an individual person in a car. They looked ridiculous, of course, and that was his point: that it is ridiculous to allow cars to take up so much of our commons space, for the sake of one person's mobility.

In one of his lectures during the Balaton Group meeting that inspired this verse, Hermann referred to sumo wrestlers — something to do with their unnecessary size, developed over months and years, all for the purpose of trying to win a very short contest. This had something to do with our cars, and the possible ways of spinning this metaphor are multiple, though I forget exactly how Hermann did it on just that occasion.

What I do recall was a series of images. He first showed what Austria looked like if you were walking (beautiful vistas), then biking (also beautiful vistas), then driving in a car (a blur of highway scenery), and finally flying (a map). As speed increases, knowledge of Austria decreases. Transiting Austria by air can be done in less than an hour, but the resulting experience of Austria is reduced to a few topographical impressions. The slower you go, the more you know, and the more pleasure you have in learning it. Our quest for speed and power in transportation has come at the cost of both knowledge and the aesthetics of human experience.

A particularly galling fact, at least from a city planning perspective, was that our unquestioning prioritization of the car made no economic sense. Downtowns do better economically when they are designed for people to *walk* around in them — not drive around in them. Even the economies of our cities would be healthier if we focused on slowing down, and increasing their beauty, not their accessibility by automobile.

As to the verse: well, it was obviously inspired by the above. Loosely.

25.

The salmon

hit the sea

and was swallowed

by a passing memory.

Commentary

This is the only verse in all of Chronosphere that features nature without people. On the surface, there is nothing uniquely human here.

Memory is a natural and necessary function shared by essentially all living things, from whales to amoeba. A sea lion accesses its memory to know where to find a good supply of fish. One could say that the memory is integral to the sea lion's identity, and fundamental to its behavior. It swims to this place, it eats this fish, because of its memory. One can also say that the memory, which is the sea lion, has eaten the fish. (A sea lion is unlikely to be the bearer of the memory alluded to specifically in this verse, since a salmon "hitting the sea" after its birth up the river would be quite small. Perhaps the passing memory in this case is owned by another fish.)

But I am being too literal. I was *not* thinking of sea lions when I composed this verse, though I *was* thinking of salmon. The measurement of wild salmon in a river can be seen as an indicator of the health of the river, as well as of the salmon, and is even connected to the health of the land around the river. In the 1990s, when this poem was written, I was known internationally as an expert on sustainability indicators. I was lead architect of a project called "Sustainable Seattle," which developed indicators of sustainability trends for that city and region, and then became a model project copied around the US and the world. The notoriety of the Sustainable Seattle project is part of what got me invited to the Balaton Group annual meeting and also what led to my international consulting career. For years, I had been traveling the world and explaining why wild salmon in local rivers — or more importantly, the *change over time* in the number of salmon in those rivers — had been selected by two hundred community leaders in Seattle as the most important indicator of regional sustainability. As a result, I thought about salmon quite often.

But I seldom thought about salmon for very long, because I was soon distracted by other thoughts. This verse is about the way our thoughts, particularly our thoughts about time, have a tendency to consume our world.

Our experience of the flow of time is a product of our observation of change in the world, and of our thoughts in relation to that world. The interplay of observation and reflection is often punctuated (or disrupted) by mental time travel: what we are observing stimulates a memory, which we replay in our minds, or a speculation about the future, which we imagine. Both memory and speculation also unfold over time, in two ways: (1) the actual time spent in the mental activity (someone could measure how many seconds or minutes we spent thinking about the memory or the imagined future scenario) and (2) the passage of time associated with the content of the memory or fantasy. That is, a memory can be recalled and re-experienced in a few seconds, but it can be *about* an event that unfolded over minutes or hours.

So perhaps I was thinking about a salmon — one of the small fry born in a frenzy of spawning that makes its way down stream and heads toward the ocean, hoping to grow into an adult fish and return one day to these same spawning grounds — when I was distracted by another thought. A memory. In a millisecond, the salmon winks out of existence, swallowed by the memory.

Just as species wink out of existence, while we are distracted with our memories and fantasies.

Or perhaps this is a construction of meaning after the fact, and the poem — as it was originally composed and intended — is just playing once again with the reification and animation of time-related concept-words in our language.

You decide.

26.

"We went to the car.

The night was fine; the jasmines

were in bloom.

We walked on by."

Commentary

Another verse about memory — this one very human, and very specific.

The quotation marks tell us that we are listening to one person talking to another (or others). Or it could be quote from a book. Perhaps it is fact, perhaps it is fiction. Of course, fictional or not, since it appears in a poem, it is also poetry.

It is in any case a story, or at least a piece of one. Telling stories is one of the oldest methods we have for communicating the events of another time — that is, for sharing knowledge with each other about the vast Chronosphere in which we live.

I read once that the fundamental ability to understand the construction *subject-verb* — *something* or *someone* does *something* — is one of the first language skills we acquire and one of the last that remains after more complex mental faculties are wiped away by age or brain damage. Subject-verb: these are the basic elements of story.

People walked. Flowers were blooming. A car stood there. The people passed it by.

If this happened, we know from the grammatical tense that it happened in the past, that it is not an imagined future scenario. We know that it happened at night. We know that it happened in modern times, because a car is involved. The word *jasmines*, hanging at the end of the line, tells us that it most likely happened in the late spring or early summer in the Northern hemisphere of our planet, that it was a mild evening, that the air was aromatic.

Most interesting: we know that a change of plans occurred. If stories are the principal way we communicate the past, plans are the most common strategy we use for shaping the future.

Walking to the car signals the closing of one activity the beginning of another, with transport in between (perhaps they are going home from an event). At least, that is clearly the original plan. That they — and the use of the word *we* tells us that there are at least two people involved, the fact that they are walking to one car means that it cannot be more than small group, and the general atmosphere of the story suggests that there are just two — spontaneously decide not to stop at the car, to keep walking, somehow inspired by the beauty of the night, all of this comes together in a few words to suggest there is an intimacy between them, and that together, they changed their plan about how to shape their immediate future together.

Their walk, and perhaps their conversation, will have continued. More than this we do not know.

But look how much we were able to read into these few lines! Or rather, how much was written into them.

27.

"Their armies swarmed across our clocks

and commandeered whole calendars,

destroyed our most precious decades.

Not a holiday was spared."

Commentary

If time is a geography, then it can also be a territory — and a target.

As in the previous verse, we are listening to a storyteller. But the singular voice, using a plural pronoun, is recounting not just a story, but *history* — a defining event in the life of a collective. The voice is explaining what happened, why things are the way they are. It is a lament: the invaders have changed everything, and much has been lost.

There is far less metaphor, and far more reality, in this verse than I realized when composing it at the close of the 20th century, before the age of social media platforms and online streaming services. These gigantic private enterprises have truly sent armies — both human and AI — to conquer ever-greater swathes of our own Chronosphere. The "war for eyeballs" they called it. To grow their companies, they needed more and more of our attention — our time — and they got it.

Combined with the relentless march of retail shopping (both physical and online), global tourism, and other forms of paid entertainment, commerce has more or less completely colonized our time. This year (writing in 2018) I noted for the first time that some "after Christmas" sales actually started on Christmas.

When writing this verse, I wished to indicate — through placing the language of time in the context of a narrative about a country — that time is also a place where we live. And like all homelands, if left undefended, our "timeland" can be overrun by hostile forces, and conquered.

28.

The Flying Woman reclined in a past-

oral scene, eating the present

offered by her future

love, Forever.

Commentary

After a long hiatus, the Flying Woman is back. In this short cinematic appearance (suggested by the word "scene", though this could also indicate a painting), she neatly summarizes all of the linear time that we humans experience.

Past, present, future — forever.

But "past" is cleverly (or stiltedly) interwoven with "pastoral", conjuring up a country landscape, perhaps a picnic by a stream, a few sheep grazing in the background. She "reclines" here, the way we rest in our past. She consumes the present as though it were being popped into her mouth, one grape at a time — each grape a "present", an offering, a gift.

These gifts arrive from the future — or rather, they *are* the future. They pop into existence, become the present, just as they are being consumed.

That the process appears endless — pop, pop, pop — is what creates the human notion of "forever", an infinite future. ("Forever" is never about the past.)

And yet, as the relative lengths of these lines also indicates, the past always seems to us to be the longest time, much longer than the present. The future, being imaginary, seems shorter still.

And forever is but a moment.

PS: Is it purposeful that the world "pastoral" is broken up so that it can also be read as "past oral"? Yes. There is more than one way to read that, as well.

29.

The tutor and his students strolled

through cherry blossoms, odd

for February. "Speak to us of time,"

they implored. He averted

his eyes. "I suspect

it is a form of ignorance," he replied.

Commentary

This event happened.

Or rather, a version of this event actually happened, to me. The event has been stylized for the sake of poetry. Details below.

The idea that time, as we humans experience it, is a "a form of ignorance" has already been established. Modern physics tells us that what we experience as time — the linear process described in the previous verse — is an illusion. Perhaps a necessary one, in that it allows our brains to manage our continued survival, but it is not to be confused with physical reality. The "tutor" in this verse appears to find this puzzling fact somewhat embarrassing.

That cherry blossoms have broken out in February suggests that something is wrong with the normal cycles of time. It is in these moments of breakdown that we are awakened to the fact that natural time is not the reliable process we wish it to be, despite our efforts to control it with clocks and calendars. In 1998, these breakdowns were odd. Today, as climate change has become manifest (more or less exactly as we feared would happen, back in the 1990s), these breakdowns are frequent events and a topic of global concern.

Finally, the verse is telling us something about how much we long to understand time. Why? Because we are trapped there.

Space is different: we can move around in space. We can change our location at will. It might require a large pile of money, but many us can, on short notice, transport ourselves to the other side of the planet in less than a day.

But we cannot transport ourselves to next year, or next Monday, or an hour from now. Nor can we revisit 1982, except mentally. We are stuck in "now". Getting to next Monday requires passing through every moment of time between now and then, in the company of every other creature on the planet. No shortcuts, no detours, not for anyone, anywhere.

Unless, in our ignorance, we are overlooking something big.

About the tutor: his name was Brendan McLaughlin, and he was my tutor at St. Catherine's College, Oxford University, in 1979-80. This scene happened essentially just as it is described, in February 1998, when I visited him at Oxford in the company of a friend (who becomes, in this stylized version, the other "student").

30.

Time is wary,

time is scary,

time is very

temporary.

Commentary

We are halfway through the Chronosphere. Time, regardless of whether it is real or illusory, whether we are afraid of it or it afraid of us, will only be with us for — yes, this is deeply ironic, but it cannot be avoided — a short time.

When our life ends, so does our experience of time. We exit the Chronosphere.

31.

When Hernán Cortéz met

the Aztec Emperor, he asked,

"How much time do you have?"

Commentary

The Aztecs believed in a prophecy: a story they had been telling themselves, for generations, about the future. In that story, their god, the "feathered serpent" Quetzalcoatl, would return to Earth. The exotic Spaniard Cortés, seated on an unknown beast and covered with an iron exoskeleton, seemed likely to be Quetzalcoatl. They welcomed him. He slaughtered them.

"How much time do you have?" is a phrase traditionally used at the start of a meeting, planned or spontaneous, where the end-time is not decided in advance. In Cortés' mouth, it sounds both humorous and evil, as though it were delivered with a leer. We, who are looking back on this event with the benefit of several hundred years of subsequent history, who know that this meeting marks the end of the Aztec empire and soon the emperor's life, also know it to be a rhetorical question.

As Cortés knows very well, the emperor has no time left at all.

My friend Wouter's cancer diagnosis was like the arrival of Hernan Cortés at the gates of Teotihuacan. Except that no one prophesied it in advance, and the conqueror was not welcomed.

32.

Anytime kept hounding Sometime for a date. "When?" she would ask.

"Oh, whenever," he would answer.

"Why not now?" "Now's as good as ever." "Ever's too late." They agreed, at last, on never.

Commentary

Anytime, sometime. The first is open to all future possibilities, the second suggests a more specific, but as-yet-undefined, future moment. At least, in theory. In practice, the phrase "let's do it sometime" often means "it is very unlikely that we will do it."

There are puns and other word-games here. "To ask for a date" can mean to invite someone out for a romantic evening, but it can also mean a request for a specified day on the calendar. Adding "ever" to "when" makes the time less specific and, depending on who is speaking and what tone of voice is employed, places the word somewhere between "anytime" and "sometime".

Why doesn't their date happen immediately? Because "now" is as good as "ever" — and "ever" is not "now". It is a strange expression, but a common one in English, to say that a specified time is "as good as ever". Usually it implies a certain reluctance on the part of the speaker. By itself, the word "ever" is a kind of positive "sometime": "Will you ever go out with me?" A "yes" answer means, "yes, sometime." I will do it, but I prefer not to say exactly when.

But add a little "n" and the word becomes "never": a time that will not happen. Or rather, an indication that there is no future moment at which a certain event will happen.

Oddly, "never" has much in common with "forever", for both stretch endlessly into the future. Infinite presence, or infinite absence.

33.

We implored the Minister

of Temporal Economics

not to trade the nation's best years

for a century of shameful mediocrity.

Commentary

In modern industrial-capitalist societies, time is transactional. It is a currency. We trade our time for money. We use money to buy time.

While there is no "Minister of Temporal Economics" anywhere in the world, there are plenty of economic concepts, policies and instruments that are intimately bound up with time: wages (payment for work-time), interest rates (payment for the use of someone else's money over time), risk (the chances that something bad will happen), and the prices of investment vehicles such as "options" and "futures" are just the most obvious examples.

But time is, surprisingly, not well integrated with economics. In 1997, when this poem was started, I spent a great deal of time reading up on the subject (because I was running an economic policy institute then, and I was also responsible for an international meeting on the topic of time and sustainability). I found very little serious, reflective research. There were time-use studies, there was minimal critique on the discount rate (the rate at which the value of a resource decreases, we will discuss this more later), and there was a book of essays on "the problem of time in economics." The problem was, economics did not really know how to treat time. In my view, it still doesn't, as searching the Internet using the phrase "economics of time" will reveal (as of the year 2020).

Thus was born the Minister of Temporal Economics. The transaction described in this verse is typical of the trade-offs that are considered optimal in standard economic theory. Exceptional things are worth less than predictable, less beautiful, less wonderful things.

And the future is always worth less than the present.

34.

She: "When a woman looks into her lover's eyes,

she sees a baby, God's eternal child."

He: "When a man in love this same inquiry tries,

he sees a goddess — ancient, wise, and wild."

Commentary

I spent a great deal of my youth immersed in a cultural phenomenon we called the "New Age". The New Age, in its North American variety, was spiritual without being religious, body- and sex-friendly, and optimistic about the power of mind and intention to shape the future. It had a few serious academic and intellectual champions — I came to know many of them — but most of its actual adherents were non- or even anti-intellectual. What the New Age movement did (for us who participated in it) was to celebrate "getting in touch with your feelings"; encourage practices like yoga, meditation, and Tai-Chi for "listening to your body"; focus attention on the self as a kind of cosmic instrument or antenna for finding a sense of meaning and purpose; and stimulate a "connection to the Earth" through healthy eating and time spent in nature.

Often criticized for being apolitical and disconnected from the real social and environmental struggles of the day, the New Age was nonetheless a pleasant way to spend one's mid-20's to early 30's. While many of the activities we engaged in are laughable now, I do not regret that time, and I vastly preferred the New Age and its flawed ideologies to the many years I spent in church as a child. Indeed, it provided a kind of healing.

For the more intellectually inclined, the New Age also invited an engagement with the universal psychology of Carl Jung, with the continuing relevance of myth in the modern world, and with the power of archetypes to shape consciousness. A professor named Joseph Campbell made headlines in the late 1980s by attracting millions of TV viewers to a six-part series of lectures where he explained these things and provided philosophical advice about how to apply such understanding in one's own life.

Perhaps most importantly, the spirit of the New Age validated those deep feelings of intuition or knowing that some of us occasionally felt, and it acknowledged the idea of mystery: that not everything that we experience has been explained by science, or can be "explained away" by it, and certainly not the mind. There were ways of knowing that went beyond sensory impressions and the analysis of data. Even my friend Wouter, a highly trained nuclear physicist, believed this, and these "subtler ways of knowing" were sometimes the subject of our conversations.

This verse, which seems to hint that the Flying Woman and Falling Man are speaking (though perhaps it is not them), captures a small piece of that earlier New Age phase of my life, which, by the time I wrote these words, was essentially part of my own past. Getting more deeply engaged with science and economics and the very real problems of climate change, racial discrimination, gender equality, and the global politics of sustainable development had diminished my interest in vague ruminations about consciousness and the power of myth. By the time I wrote Chronosphere, I had switched from yoga to martial arts, and started eating meat again.

I would have likely composed these lines with an ambivalent mixture of self-deprecating irony ("I can't believe I used to believe that stuff") and an appreciation for the truth of our archetypal patterns, especially when it comes to traditional, heterosexual gender roles. Gruff masculinity often masks a tender boy, who can be brought out to the surface by a caring lover. And every woman bears in her body the mystery of life and birth, which was the basis of many goddess cults and religions in the ancient world, and which still has the power to mesmerize a man.

While Campbell's interpretations would need to be updated for a world where non-traditional gender roles and other forms of sexuality are now mainstreamed, there is a central and inescapable truth in these neo-Jungian reflections: on our passage from birth to death we must negotiate our relationship to these roles and archetypes and their roots in both our bodies and our cultural heritage. Deep time, ancient history, is still with us in the present moment.

35.

The Falling Man hummed

a morning raga

all afternoon.

In the evening, the sun

bathed the Ganges

in robes of saffron.

Commentary

When studying Asian Philosophy at Tulane University in New Orleans, I was taught that Asian cultures had a different concept of time. In caricature, "they" were cyclical, "we" were linear. The Western world had more or less invented the idea of progress, said the professor, whereas the East was stuck in the notion that what goes around, comes around, endlessly.

Later I was introduced to the patterns of Asian music, with its simpler scales, rhythms and compositions — at least, this was how it was presented to me, for it was framed in sharp contrast to the monuments of the Western symphony orchestra. It made sense that one of the Beatles would play sitar with Ravi Shankar, or possibly a jazz musician, in this framing, but the hierarchy of accomplishment and complexity was thus left undisturbed.

Perhaps because of the arrogance implicit in these presentations, I found myself drawn to this Otherness. Offered a choice between returning to Oxford for more studies or competing for a fellowship to live and work in Asia for a year, I chose the latter, and won.

A year of living in Asia, in extreme conditions, thoroughly overturned the default conceptualizations that framed my exceedingly Western education, but not in the ways that I expected, because my expectations had been framed by that selfsame biased, arrogant, and deeply uninformed education.

The principal philosophical systems that originated in Asia (represented today by words like Hinduism, Buddhism, Taoism) are all significantly more advanced than their western counterparts in their treatment of life, death, and especially consciousness. Western philosophy ultimately led to successful systems for understanding how the world works. Eastern philosophy led to systems for understanding how our minds work.

I suppose I expected nirvana from my year of living in Asia. Instead, I learned that life is inherently chaotic, that people tend to be wildly irrational (everywhere in the world, certainly not just in Asia), and that abstract thought does not prepare us to navigate these things. Only experience does.

But still I — and this verse — return to the old, stereotypic trope of Asian serenity, which also partakes of truth. The Falling Man is preparing for the end by singing the calm, ritual music of the start of the day. The timeless image of the Ganges, whose illumination by the sun evokes the raiment of the lifelong contemplative, suggests that he is at peace with the fact that he is dying.

This is only a passing notion.

36.

In the cafe of history,

a pair of important events

sipped espresso

and spoke of mutual

episodes.

Commentary

An aside: I love cafés. They are one of the few perfect places to spend time, landing, for me, on a kind of mental top-ten list that includes things like "in my lover's arms" and "sitting by a beautiful river". For these are places where the amount of time one spends hardly matters: ten minutes, ten hours, none of it, short or long, is ever regretted. (There is, of course, an order of preference, and cafés are not number one, but they rank highly.)

This verse is itself an aside, another parallelism in the narrative, which could be described as fractured and full of "meanwhiles". In these reports from elsewhen, temporal concepts are anthropomorphized, personified, and placed on a stage. In this scene, even the totality of past human experience that we call "history" has its hole-in-the-wall, out-of-the-way little locale where folks can pause and chat about halfway meaningful things. And sip a coffee.

The folks in this café are "events", and they have status. They are the kinds of events that end up in history's hall of fame. They contain multiple timelines within them, discrete smaller parts, often repetitive or predictably similar. The nation goes to war and the king dies, a scientific discovery upends the old order, a religion emerges and conquers half the world. Events are themes, the bolder strokes of history. Within them, episodes: there might be a side-story of a jealous brother, an illicit love affair, a woman who bucks oppression and does what she wants, and most certainly needs, to do.

"Really," nation-goes-to-war might say to religion-emerges, nibbling on a biscotti, "that happened with you, too?"

37.

A messenger approached the throne.

The Flying Woman leaned over,

her face composed of stone.

"What news?" "Their barricade

is broken. They cannot long

sustain themselves. We've won."

She stared into the day.

"They must feel so alone."

Commentary

The moment of victory, like all moments, is not isolated in time or space. The Flying Woman tastes the bitterness of her foes' experience — the inevitable unfolding of defeat — before it has even occurred.

But what defeat? And who are "They"?

Here we have a significant clue to the identity of the Flying Woman. She is seated on a throne. She is at war, or is at least attempting to crush the defenses of another party.

Just as time inevitably crushes our defenses. (See verse 27.)

They is we.

38.

Three cheers for the Falling Man!

He's tempirical! He's

mentaphysical! And when

the times get tough, he falls

to the occasion!

Commentary

Hurrah! At this point in our meta-narrative, the hero needs a bit of cheering up, and so do we. Facing the ultimate defeat — the end of life — is not easy. My friend Wouter was tough. He taught me many valuable things, but by far the most precious was how to greet death with bravery.

The Falling Man faces his diminishing stock of time with a sense of empirical clarity ("tempirical"). He overcomes the slow wasting away of his body by setting his mind to work during the time still remaining to him ("mentaphysical"). And when the final decline comes, he accepts it, "rising to the occasion" by having the grace to go down, without self-pity.

So let us celebrate the nobility of the Falling Man. Hurrah, hurrah, hurrah!

39.

Down with the Falling Man!

He screws up! He's lost

the womandate of heaven!

He can't tell a sphynx

from a sphincter! He

proves the theory that you

can't keep a down man good!

Commentary

Hapless. That is what the ineluctability of dying also makes us into: hapless idiots who refuse to understand just how screwed we really are.

Let us be a bit blunt here: colon cancer is no fun. You cannot deal with your own shit. Your shit deals with you. We are used to practicing control over everything having to do with our anal regions; that is what my generation's blithe references to people being "anal" (a sort of Freudian code word, meaning extreme in their need for order and control) are all about.

There is no deep mystery in the line about sphynx and sphincter, at least in the poetic sense. The latter is the muscle that our parents are most keen for us to learn to control, so that we can deal with our own shit and they can stop changing our diapers. The former is a common symbol for riddles and mysteries. When we, at the end of life, for reasons having to do with illness or old age, can no longer practice the fundamental muscle control that allows us to deposit at will our shit in preordained receptacles called toilets, we are confronted once again with the mystery of what it means to be alive, and why our bodies must ultimately break down and die.

We can also no longer participate in one of the fundamental rituals and systems of civilization: sewage management. We are helpless, and therefore hapless, and feel deserving of disdain.

Or at least, so I imagine the feeling to be. There is an embarrassment associated with that haplessness, no matter how loving and accepting other people may be about one's increasingly wretched condition. Why, oh why, has the universe forsaken me? The Falling Man, so recently cheered and celebrated for his bravery, is now experiencing a moment of crushing existential and social failure. He no longer feels "good".

Take just a few more wobbly paces down the path of corporal decay, and even the capacity for reveling in the mysteries of the spheres, sphynxes, or sphincters also dissolves into some simpler level of consciousness. Ultimately, we can no longer tell the difference between the deepest philosophical mysteries and the most humiliating physical miseries. If we are lucky, we stop caring, or noticing.

Let us also unpack a few other playful phrases here. Why, when we make great errors, do we screw "up"? I do not know, but this works well in contrast with falling down. Emperors in China were said to lose the "mandate of heaven" after screwing up (or when they were getting blamed for societal calamity). To add "wo" gives another small link to the Flying Woman and her divine justice. And finally, the Falling Man's imminent demise is living (or dying) proof of the essential theoretical starting point for this poem: life is damned short.

40.

Time is funny,

time's amused,

time has nothing

much to lose.

Commentary

Active, passive. Emptiness, fullness.

You can scan those four words against the four lines of this quatrain, and the associations will make sense, even if the verse is best read by a court jester in a funny hat.

Alternatively, we might imagine that Time is a criminal, a character in a Raymond Chandler novel. The film version would be shot in black and white, and Humphrey Bogart, playing Sam Spade, would be describing Time to a police officer. "What can you tell me about Time?" the officer would say. "Well," Bogart would reluctantly avow, "Time's funny, and I don't mean funny ha-ha. Time also seems to be easily amused, judging by the trail of not-so-funny corpses left in its wake. But the truly dangerous thing is" — and here there would be a drag on a cigarette, an exhale of smoke — "Time just has nothing much to lose."

Because time is a kind of nothingness. We cannot see it, we cannot stop it, we cannot affect it. And yet everything we know and love (and hate) is locked within its obvious cycles and spirals, peaks and abysses, dashes and dots.

Actually, it is we who, living in time, have much to lose.

41.

The Falling Man

took his time.

He felt bored.

He gave it back.

Commentary

Bad puns are no different in kind than clever metaphors or *jeux de mots*. Used purposefully, they make a blunt point with a double meaning. Or try to.

At this point in our narrative, the Falling Man has already been through hell. He endures small moments of hell every day. He is "taking his time" to fall, because the progress of his condition is gradual. He "takes his time", that is he grabs at it, because he is doing all he can to stretch that progression out, to make life last longer — despite the growing pain.

But there is also a monotony in the process of resisting. And at this moment, the Falling Man has grown tired of it. He stops trying to grab for something extra.

This is the moment of acceptance of his fate.

42.

When the still was going

full bore, boiling up waste

memory into fine hard

liquor, the smell of

remorse was overwhelming.

Commentary

Regret and remorse are emotions born of the human capacity to both remember the past and to imagine alternative futures. Researchers call this *chronosthesia* — more colloquially, "mental time travel." Our brains can both reconstruct events that have already occurred and, with nearly the same level of subjective experiential detail, construct mental pictures (and smells and tastes and feelings) of events that have not yet occurred.

You can test this yourself very easily. Pick an event in your past. Travel into it, mentally. There you are: the air felt just so, the light was of a certain quality. You can hear the voices, see the action play out, more or less exactly as it happened, at least in broad brushstrokes. Yes?

Now do the same with a future fantasy, something that you hope might someday happen. Imagine the scents, the sensations, details of the location, the people, facial expressions.

Not hard, is it? And the two experiences are surprisingly similar, mentally.

This capacity for mental time travel extends to events that *might* have occurred had we been luckier or wiser in our choices but did not. Expressed negatively, one can *imagine away* certain events from the past, and then travel down the resulting alternative timelines in the mind. "If only I had not done *that*, then *this* would not be happening." Something else would be happening instead, likely something preferable to the current reality (though of course one never knows).

We can bring these unrealized, unrealizable possibilities to mind with exquisite clarity. We can take these nonexistent yet easy-to-imagine experiences and turn them over in our mind, reflect on them from multiple perspectives — and respond emotionally to them. In some cases, the awareness of what we have caused to happen, failed to grasp, or simply missed (as one misses a train leaving the station) can bring up a deep and powerful sense of guilt and loss. This is the origin of remorse.

We know that most of our experiences are lost to us. Memories, it seems, are often stored but never retrieved. The brain is continuously prioritizing memories, partly based on how often they are accessed by the searchlight of our conscious minds but mostly on unconscious processes. Will (the person whose brain this is) need this memory in the future? If the judgement is "probably not," it is demoted, thus freeing up space and mental processing power for replaying one's highlights and regrets.

But buried in all of those waste memories must be thousands of small decisions that set our day, our year, the rest of our life on a less-than-optimal course. Fortunately, we are not aware of this, for the most part. But there must be tons — or the qualia equivalent of tons — of latent, untapped remorse in those memories, which this imaginary still in verse 42 is refining into a liquor of forgetfulness.

43.

The Falling Man dropped into

the time-farmer's market.

He touched the fresh epochs

and the tempo beans.

He tried to sell himself,

but his time wasn't ripe.

Commentary

The puns in Chronosphere are always playing with fire when it comes to the fine balance between simply being "bad" (from the perspective of humor) or being clever. Do they cause a small hammer to knock the brain a bit, by bringing everyday turns of phrase — such as "the time was ripe" — under the harsh lamplight of reflection? *Now that's clever.* Or does one just hear the snare drum signal after a dumb joke? *Da-DUM-dum. Groan.*

You decide.

What does it mean for time to be "ripe", a word we use to describe fruits and vegetables that are ready to be eaten? As a metaphor, it plays havoc with our traditional understanding of time, as something that we are *in* (like a river). Suddenly time is a *quality* that resides in events. It develops and then suddenly exhibits itself, becomes a kind of signal, as when cherries turn deep red on the tree. It is only then we should eat them.

> *When should I ask my lover to marry me? When should I change jobs? When should I launch the hostile takeover bid of the company in which I am a minority shareholder? Not now: the time is not yet ripe.*

Our protagonist, the Falling Man, is encountering these qualities in the market, just as we encounter them in our everyday experience of "the right time": the quality of time becomes reified, embodied, and consumable, but only at the optimal moment. Perhaps he is attempting a desperate short-circuit of his inevitable fate by "selling himself" in the time-farmer's market — selling his own reified, embodied, consumable time. But the pros at the market aren't buying: they know *good timing* when they see it.

The Falling Man must continue his journey. Continue falling, and waiting.

44.

The Flying Woman took a few

turns around the clock tower,

speeding up, slowing down,

doing barrel rolls above

the bustling town.

Her gyrations cast

deep shadows on the ground.

Commentary

One summer, many years after I wrote *Chronosphere*, my family and I were making our mostly-annual visit to loved ones in eastern North Carolina, in the United States. While driving across a bridge we took notice of a yellow airplane, a couple of miles away and flying low to the ground. After coming off the bridge, we pulled over and stopped at the side of the road and just watched, because it started to do absolutely amazing things.

The small yellow plane was performing just what the Flying Woman is doing in this verse: swoops, dips, barrel rolls, a kind of death-defying display of aerial prowess. Sometimes the plane would fly straight up, stall on purpose, then tumble and plunge below the distant treeline. We half expected to see a ball of orange fire erupt upwards. Never happened. The pilot was far too skillful for that. Instead, we learned to scan the horizon after one of those maneuvers, the same way we scanned the water's surface after a dolphin sighting: we knew it had to come up somewhere, even though the exact spot was unpredictable. And indeed, it always did reappear.

We learned later that these private antics of flying happened once in a while, but no one seemed to know anything else. We never learned anything about the identity of the pilot, apart from witnessing this display of extraordinary skill, exercised on a personal whim.

In the limited poetic universe that is *Chronosphere*, the Flying Woman is constantly toying with us in similarly spectacular fashion. We cannot know her. We can only observe her antics, and her power, whenever she deigns to show it. And we suspect that this power is not ultimately deployed to benefit, nor even to entertain, the people of the bustling town. Or the Falling Man. Or us.

But we watch nonetheless, transfixed. Wondering what comes next. And when.

45.

Hunger gripped the Falling Man

like a tourniquet. He had eaten

all the sweetness of expectation,

gorged on expanding the present

moment. He nibbled fantasies and

memories — they were mere

distractions. Without a steady diet of

experience, he shriveled like a wraith.

He needed something to bite into;

there was nothing but bitter faith.

Commentary

The poem *Chronosphere* continues to develop time-as-food metaphors. In verse 43 time was embedded in the ingredients offered at a market; in verse 45, time is the meal, and it's all about eating.

But what's happening to the Falling Man? Now he is ravenous. He gorges and devours, but he is still hungry. Everything he *can* eat is unsatisfying. Expectations, memories, even the all-encompassing now-ness one seeks in meditation are mere fluff — internal subjective states lacking in external substance — compared to what he really hungers for, and needs: more experience, more life.

To experience something is be in direct sensual contact with it. Of course, we talk about "internal experiences", like thoughts and dreams, but we also generally agree that there is a fundamental difference between these and real-life experience. People trapped in coma are apparently having experiences, but we would not say that they are having *life* experiences, because they do not have the option of involvement and agency in relation to something outside themselves. They cannot change the world or be changed by it.

At death, so far as we know, all possibility for true experience of this kind simply ceases. We can no longer participate in the steady flow of events, one thing happening after another, that gives rise to the metaphor of time-as-river. The great and melancholy burden of being conscious in all the ways that *Chronosphere* tries to portray, with its poetry and puns and playful imagery, is the inescapable knowledge of this ultimate ceasing — not of the world, but of our ability to participate in it. Religions offer some salve with which to treat this knowledge, but no cure: hence the bitterness inherent in faith.

Which brings us to the question of life after death.

As a very young man I read a book on near-death experiences that shook me to the core. The first-person testimony about these experiences was surprisingly consistent: traveling down tunnels towards a light, being surrounded by departed loved ones, feeling greatly at peace, then snapping back into a body that had suddenly begun to recover from accident, trauma, illness or surgery. These descriptions had a very different qualitative feeling in comparison to the fantasies, memories and dreams about which I have heretofore been writing.

In a word, they felt real.

This feeling of realness was strengthened by three things: first my mother and later a dear and wise friend both reported experiencing exactly what the book described, at moments of critical illness, after which doctors had noted that they could just as easily have died. Their subsequent (and surprisingly swift) recoveries were not classed as miraculous, just fortuitous. They had gotten very close to the brink of death, peeked over the edge, and come back. Both of these loved ones, people I knew never to lie or to fabricate, were simply astonished at what they saw and heard "on the other side". For them, these were true experiences: they had, at the point of near-death, been in direct sensory contact with things and processes that were objective and outside themselves, not the products of their own conscious or subconscious minds. They were as convinced of the reality of their experience as they were of their memories from last Wednesday, perhaps more so.

The third was a similar experience of my own that was not, however, caused by illness or accident. Nor was it the product of hallucinatory drugs (which I have never used) or any other external stimulus. It was something that just happened to me during my university years, after an intensive period of philosophical reflection. It was unexplainable. It felt utterly real. And frankly rather scary. When it was over, I felt both relieved and bereft: I had glimpsed something beyond myself and this world, something I desperately wanted to know and understand. But I was given the impression that full understanding would have cost me my life, in some sense of that phrase. So I had closed that door firmly. I was haunted by a sense of loss for years afterward. But I was also happy to be alive.

To say that all these experiences *felt* real to the individuals who had them and reported on them is not the same as *knowing* that they were real, for such things cannot be verified by another person (which we can allow to stand as a rough definition for "objective"), not even theoretically. In later years, informed by reading a lot of neuroscience, I have come to regard experiences like these with a great deal more skepticism, but I still cannot dismiss them either. All three of us were convinced that these things had "happened", in some real sense that was distinct from other subjective, inner experience.

Do experiences like these offer any proof of life after death? Not in the least. But they do offer the strongest possible evidence for the fact that to be alive, as a human being gifted with self-reflective consciousness, is to hunger for new experience — and that this hunger follows us all the way to death's door, and perhaps beyond.

46.

The poet raced against time

to complete a poem about time

for a man whose time

was running out.

Commentary

The concept of a "deadline" has never felt so tangible to me as when I was writing *Chronosphere*.

My friend's treatment options had been exhausted. He was starting palliative care. I hoped he would survive until our planned last visit, and that I would finish my work *in time*. Unlike the concepts of respite and reaction times that Wouter had introduced me to, the completion of my poem, whether in time or not, would have absolutely no impact on the outcome of the processes set in motion by his cancer. The only difference was whether he would *know* that I had written a poem inspired by him, or not.

And what difference would that make? Objectively, none. But at the time, beating that deadline was a matter of utmost importance to me. A matter of "life and death."

There was a literal stripe across the road of time, a real deadline, somewhere ahead of us. Once Wouter crossed it, that was that. But there was no telling exactly *where,* or rather *when,* the line in time actually was drawn. That sort of deadline is utterly and irrefutably real, and yet we cannot say anything precise about it until we suddenly (and unpredictably) arrive there and certain events actually and irretrievably occur. It is something like Heisenberg's uncertainty principle: the position or trajectory of a particle is all a matter of probability, and quite unknowable, until it is measured.

The Chronosphere Commentary

And death is the final measurement.

47.

The conductor set the metronome

to one beat per month. Entire

generations of musicians

spent their lives rehearsing

Brahms' Fourth Symphony.

Commentary

Nowhere is the measurement of time more essential to the essence of a thing as it is in music. Slow, fast, regular or varied, salsa or raga or symphonic march, music fills and demarcates time in beats, measures, *tempo*. Musical compositions have a wonderfully named element in their fundamental construction that tells all future musicians how to organize the playing of the notes in temporal terms. It is called a *time signature* — as though time itself must sign off on every song, chant or aria we create.

This verse of Chronosphere, which imagines what would happen if the tempo of a piece of music were reduced drastically (by a factor of several million times slower than normal), is not as fantastical as you might imagine. In fact, it was predictive. The idea of extremely slow music, played over generations, is now a reality. In a small church in Haberstadt, Germany, a composition by John Cage called Organ^2/ASLSP (As Slow as Possible) began to be performed in 2001 and will not be completed, if all goes as planned, before the year 2640. (Chronosphere was composed in 1997-1998.) Sandbags placed on the organ pedals allow for certain chords to be played over the course of months and years before they need to be changed. The initiators of the project have committed over 30 generations of future musicians to participate in this singular performance.

A 640-year version of *ASLSP* is an act intergenerational faith. But so is civilization. Many aspects of human life are organized around the belief, or at least the fervent hope, that others will take on the responsibility of continuing various processes that we have either started or inherited from our forebears. Religions are the most long-lasting examples we have: Buddhism, Hinduism, Jainism, Judaism and Zoroastrianism are each more than two thousand years old. But the practice of science (including published and verifiable evidence), universities as institutions, many nation-states, and the cultural traditions of literature written in a hundred different languages also have histories of continuity that are measured in centuries, coupled with the expectation that others will carry on doing the necessary activities for hundreds or even thousands of years to come.

In our era, science also has been a source of worry that humanity's dominant sense of time is too short. As religion and the nation-state have receded in their importance, to be replaced by the faster-paced advance of technology and market-based economic processes, the mass of humanity is decreasingly focused on inter-generational duty — just when our understanding of our long-term responsibilities has increased dramatically.

We now know that maintaining both human civilization and a viable biosphere on planet Earth requires thinking, planning, and concerted action over very long time scales. We tend to plan in annual cycles, a decade at most. But decisions made today about how we power our cars, planes, homes and factories will have thousand-year implications. And decisions that end up erasing natural habitats or disturbing entire ecosystems can put an end to evolutionary stories and environmental balancing acts that were millions of years in the making.

Expanding our sense of time is one of the core requirements of *sustainability*, the concept and practice to which my friend Wouter devoted his professional life. Time sense is just as central to sustainability as it is to music (I happen to practice both). It is of course one of the reasons I wrote Chronosphere. And it is also one of the principle motivations behind a marvelous project called the Clock of the Long Now, a real, mechanical clock designed to tick for 10,000 years without winding, now under construction inside a mountain in the western part of Texas, USA.

Here is what the clock's inventor, Danny Hillis, has to say about the project:

> *I cannot imagine the future, but I care about it. I know I am a part of a story that starts long before I can remember and continues long beyond when anyone will remember me. I sense that I am alive at a time of important change, and I feel a responsibility to make sure that the change comes out well. I plant my acorns knowing that I will never live to harvest the oaks.*
>
> *I want to build a clock that ticks once a year. The century hand advances once every 100 years, and the cuckoo comes out on the millennium. I want the cuckoo to come out every millennium for the next 10,000 years.*

The common assumption or hope behind the Clock of the Long Now in Texas, the multi-century *ASLSP* performance in Haberstadt, and verse 47 of *Chronosphere* is this: if we can better understand long timescales, we can better manage human civilization.

48.

The Falling Man made a graph

of his descent. The curve

was the shape of an epitaph.

Commentary

One of the first things I learned about in the informal School of System Dynamics that I attended annually on the shore of Lake Balaton, where I first met Wouter and many other systems experts and learners, was how to construct a diagram called a "behavior-over-time graph" – BOT for short.

I had first encountered BOTs in the classic book *Limits to Growth*, written by the founders of the Balaton Group, Donella and Dennis Meadows in 1972. *Limits* attempted to show the world what would happen if we kept on growing, using up resources and spitting out wastes in the manner we had already become accustomed to in the 1960s and 70s. The output of their computer models, presented as graphs of what was most likely to happen to the world's economic output and population over time, inevitably showed a peak in human welfare somewhere around the early-to-mid 2000s, followed by a crash back to levels roughly equivalent to those of the 1920s.

In my own first book, *Believing Cassandra*, I described the lines on those graphs (scientists prefer to call the lines on graphs "curves", even when the lines are straight) as "swan dives". Humanity leaps up, reaches a beautiful apex, then falls headfirst into dystopia. Coded into the graph is the grim awareness that many people have to die along the way, at least in theory.

The world is still waiting to find out whether the authors of *Limits* and other scientists will turn out to be truly successful prophets — meaning the kinds who manage to warn people away from such likely-but-avoidable negative outcomes, thereby proving themselves "wrong" — or merely accurate ones, whose foretellings come true, to the consternation of the prophets themselves and to the world's deep dismay.

But of course, every person's life can be graphed in this way. The word "behavior" in BOT does not mean the things we do; it means the evolving state of any system that one happens to be looking at, as measured by some indicator. And people are systems.

Let's say the indicator is some combination of health, prosperity and individual subjective wellbeing — how happy we are, during the course of our lives. At birth, we have little and know nothing about our own lives. During the course of a happy life, the life-satisfaction BOT-graph would trend upward. A relatively sudden death would just be a straight line down to the X-axis: that is the "deadline" considered in verse 46.

But for a person facing a chronic and then terminal illness, the curve might appear more like those in the book *Limits to Growth*, with an apex followed by a gentle downturn (at the moment of diagnosis), and eventually a steeper descent. One would have time, during that descent, to compose a few lines by which one wished to be remembered.

"Words upon a tomb." The definition of an epitaph.

49.

The Falling Man's research

revealed an imbalance

in temporal trade: those

in wealthy countries

purchased extra years

from the poor, consumed

twice their share of time,

exported their despair,

delayed their inevitable

deaths by living lives

that might have once belonged

to, say, a young mother,

shaking off flies in the Sahel.

Commentary

In a rather profound way, this verse needs no interpretation. After the third line, it becomes a despairingly accurate description of the global situation. We generally talk of poverty and global inequities in terms of money, purchasing power, access to resources and education. But time is the great equalizer. Or unequalizer.

If you were born in one of the world's poorest nations, your life expectancy at birth was between 50 and 60 years. In the richest nations that number has now crept up to over 80 – a 20 to 30-year difference in how much life you can expect to enjoy. Or suffer, depending on your circumstances.

This profound aspect of inequity is usually ascribed to differences in national and individual income, and then passed over, as though wealth (or poverty) were some kind of birthright. But what about trade? What about the many dangerous or difficult jobs — jobs that take years off one's life — that have been shifted from rich countries to poor ones?

Faced with the knowledge that we who live in the wealthier parts of our world are consuming other people's lives, and thereby living not on "borrowed time" but actually stolen time, what should we do?

50.

Time is pricey,

time is free,

time is nothing

much to see.

Commentary

Much of our daily relationship to time is hidden under layers of concepts derived from the dominant discipline of our age: economics.

We sell our time in the labor market: this is called a "getting a job". We buy other people's time directly via the service industries, and indirectly whenever we purchase a product. But that is just the beginning of how economics has swept up and obscured our relationship to time to such an extent that we can rarely see it, much less question it.

Consider the commonly used but little-known economic concept of the "discount rate". This is the rate at which the value of something in the future, an asset or an outcome or even a human life, declines over time. Simple example: suppose you own a forest. When calculating its value in ten years, predicted changes in the price of timber (assuming you want to sell the forest) must be offset by the risk that your forest will not be there ten years from now. It could have burned down, for example. This gives you an incentive to sell it sooner rather than later: you should realize your profits before it is too late.

To summarize grossly, when economists look into the future, they see a world of resources that are worth less and less over time, because of the steadily increasing risk that things might be spoiled or destroyed or made unsellable for other reasons. Here is how Wikipedia explains the key assumption underlying this remarkable concept:

> *Since early in the twentieth century, economists have analyzed intertemporal decisions using the discounted utility model, which assumes that people evaluate the pleasures and pains resulting from a decision in much the same way that financial markets evaluate losses and gains, exponentially 'discounting' the value of outcomes according to how delayed they are in time.*

Do you agree with this assumption? Are your decisions about whether to get married, have or adopt a child, care for a piece of land, write a poem, etc. etc. governed by the same logic that drives a broker (human or robotic) who is considering the purchase of a thousand shares of stock? Do you look at a piece of rainforest and think, "this will likely be much less valuable a hundred years from now"? Because that is what current economic theory — the "invisible hand" currently steering much of our world via concepts like these, built into a thousand equations on which decision-makers rely — assumes that we all believe and accept.

When verse 50 says "Time is pricey", it is not just saying that time can be expensive. It is saying that *time, in all its human-experienced forms, tends increasingly to have a price attached to it*. Whether we like it or not, "Time" (wrote Benjamin Franklin in 1748) "is money."

51.

A monk and a capitalist talk

in Thailand. They watch the waters

rise. "Tell me the value of

breath," says one. The other sighs.

Commentary

One might assume that the capitalist is the one asking the monk and that the monk is sighing to indicate that the capitalist's question is typically predictable. But perhaps the capitalist is mocking the monk for spending inordinate amounts of time observing the process of breathing instead of generating value in the market economy. Or perhaps the capitalist's question is genuine and guileless, and the monk's answer is both non-verbal and non-ironic: to sigh is to exhale. Breath is breath. Leave aside concepts of all kinds, the monk could be saying, including value.

But the verse does not specify who is asking whom. The monk could also be interrogating the capitalist, trying to root out unexamined assumptions or to point out that not everything can be monetized and priced. Or perhaps the question is meant to arrest the capitalist's process of thought, in the way of a Zen *koan*, and the capitalist's sigh is an outward expression of the inner realization that chasing money has been an empty dream. Or perhaps the monk's question is genuine and guileless, and the capitalist's answer is both non-verbal and non-ironic: "Even I know this. Breath is its own value."

But why are we in Thailand? And where is time in this exchange? It is in the water. A flood is coming. Thoughts can be stopped, but not the accumulation of rain in the river.

This verse was partly inspired by the stories and the personality of my friend Chirapol Sintunawa, a member of the Balaton

Group who has spent time as a monk but is principally a professor (now emeritus). Chirapol could easily have told a story like this one, for he talked to us often about finding new ways to convey the combination of both systems and spiritual insights to people in the mainstream of business and politics in Thailand. Chirapol became relatively well known in his home country, advising the former King and many others on environmental matters. I seem to recall his coming to a Balaton Group meeting, perhaps on the subject of water, and talking about how to use systems thinking to prevent floods; and perhaps this verse popped out of that context — if not actually, then theoretically.

Wouter, who knew Chirapol well and whose interests were wide-ranging, would have recognized all of the above aspects of this verse. Being terminally ill, he would likely have had somewhat different reflections on the value of breath.

52.

The Falling Man went to a barbecue

in Botswana. They cooked

his watch. The minutes

tasted like hours.

Commentary

Years after writing Chronosphere, I visited Botswana and numerous other African countries in the course of my work. To say that African cultures have a different sense of time from most European or American ones (those are my home cultures) is like saying that apples are different from oranges: the statement is true, but meaningless.

All cultures experience time differently, though the differences can be more or less extreme. As I traveled and worked throughout eastern and southern Africa, what proved truer than I expected about verse 52 — even after a lifetime of globetrotting and living in different parts of the world and experiencing many concepts of time — was the "cooking" metaphor. As a "short wait" of what I presumed to be five minutes sometimes stretched to five hours, I began to see that immersing oneself in the reality of another culture is not, in the first instance, like wandering through a market, or putting on a different set of clothes. It is more like turning your habits and expectations over to the kitchen staff of a previously unvisited restaurant, as raw ingredients, and having those concepts returned to you on a plate in a transformed state, tasting wonderfully different. Cooked.

The Falling Man has been all the while teaching us about time through his own falling and learning. Time is priceless. Timescales can stretch or contract. Subjective experiences of time can disagree with objective measurements. Even the words we use to describe our experience of time are arbitrary constructions:

where is it inscribed in the heavens that "minutes" or "hours" even exist? They are not like *pi*, the fixed number that describes the relationship between every circle in the universe and its radius. *Pi* enjoys a kind of independent existence, regardless of what we call it ("a rose by any other name would smell as sweet" wrote Shakespeare in *Romeo and Juliet*, published 1597 — that is, 423 solar orbits prior to my writing these words). Minutes and hours, in stark contrast, are just concepts invented by the ancient Babylonians. They are tools that divide our collective, physical experience of time's passage — the sun rising and setting, the moon waxing and waning, the seasons cycling as the planet parades around the sun — into smaller, measurable, communicable pieces. We could just as easily be living in a culture whose language lacks any words for "minute" or "hour", and there have been many such cultures. "In five minutes" might be translated the same way as "in five hours" — with a single word that we might call "soon".

But the modern world depends absolutely on the precise measurement of time. Measurement depends, in turn, on precise definitions. Here is the definition of one second, as agreed in the International System of Units: "the time it takes a Cesium-133 atom at the ground state to oscillate exactly 9,192,631,770 times." That is a very specific but exceedingly arbitrary number.

The time we measure and experience in the industrialized world, whether in Africa or Antarctica or anywhere else, is not *itself* arbitrary, for it is based on real things happening in our universe. But our *approach* to dividing time up, measuring it, and talking about it, most decidedly is.

53.

The Falling Man catalogued

his accomplishments. The list

was very long. He was over-

come with joy, and broke

into poetry, canticle, and song.

Commentary

I do not remember any of my father's words of advice to me as a young man except for this one sentence, which he repeated often: "Everything in life is about how you spend your time." He meant that if you spend it productively, you accomplish things. And by "accomplishments" he meant the kinds of things one might list on a *curriculum vitae*.

Let us leave aside the question of whether "accomplishment" of this type is something worth striving for in life. In simpler terms, most of us just want not to feel that our lives have been wasted. Or pointless. Instead, let this word "accomplishments" stand for all the things one has done — or been permitted to do — over and above the minimal requirements for survival, during the limited time afforded to us on this planet.

For many people, life is not about accomplishment because it cannot be. It is about the fundamentals of feeding one's family, sustaining a home, and dealing with the never-ending challenges of poverty or oppression. The notion that one might get a higher education, pursue a meaningful career of one's own choosing, or in one's free time conceive of projects, born of ambition or vision or hard-to-describe inner urges, and thereby generate a list of memorable deeds has only become possible for the middle-class-mainstream of global society during the past hundred years or so. Prior to that, a focus on achievement and accomplishment was a luxury limited to the relatively wealthy or to those who could wrangle some of their support, through various kinds of patronage.

But today, the majority of humanity does have the luxury of choosing work they might actually want to do. Most people have, at the very least, some personal time to spend on non-essential activities. The invention of weekends and holidays (fought for by the labor movements) and the general increase in productivity made possible by modern energy and technology have given billions of people the opportunity to ask themselves, "What would I really like to do with my free time?"

In the statistical databases of the modern industrial societies, this aspect of "time use" is usually measured under the category "leisure". The amount of time spent on "leisure" varies from country to country, and even between men and women (men generally enjoy more of it). Much of that leisure time is spent passively, watching other people doing things. Vast industries — television and streaming, pop music, professional sports, the list is endless — have grown up around people's availability to do passive things of their own choosing, outside of worktime and the time needed to meet basic needs like food and sleep.

But another, more impressive result of this general transformation in how humanity uses its time is that the list of our collective accomplishments is expanding at an astonishing, accelerating rate. From self-published poetry collections to academic papers on cosmology to TikTok videos, we are, in stark contrast to previous generations of human beings, completely flooded with the productive output of our own professional and leisure-driven creativity.

And yet, here is a sobering fact: in the European Union, only a quarter of the population say they are "highly satisfied" with the way they spend their time. Half are just "moderately" satisfied, the rest report low satisfaction. And that is in the EU, which includes many of the world's richest countries, with the most generous policies around time free from the obligations of work.

In other words, most people in this world would probably not break into "poetry, canticle, and song" if confronted with a catalog of their personal accomplishments and the news that their lives were soon finished.

Our Falling Man is therefore exceptional. True, he is approaching the end of his fall (there are just eight verses left in Chronosphere, the deadline is fast approaching). But he can look back on his life with a very high sense of satisfaction, which clearly lifts his spirits and mollifies his melancholy.

What is it that has made him happy?

What list of accomplishments would make *you* happy, in similar end-of-life circumstances?

And how do we build a world where *everyone* has the possibility to choose how they spend their time?

54.

"There is too much of everything

but time — too much birth, too much

death, too much of the stuff

in between, both the done

and the seen. O let life

be one day longer than it seems."

Commentary

Who is speaking here? The speaker is unspecified, and I cannot myself recall exactly what I was thinking at the time. But there is a hint to the identity in both the previous and the following verses.

This is very likely the Falling Man. He is very likely singing. And this is the lyric to his song.

Something has happened to the Falling Man. His myriad experiences and encounters during the long fall have transformed him. He embraces what is happening. From here on out, he is not a victim, though he retains his melancholy. He seems to be caught up in a kind of creative ecstasy. He is making his final mark upon the universe, while also lodging a complaint.

However long life is, it is not long enough.

55.

The Singing Falling Man

attracted the attention

of acoustical astronomers

who Dopplered and Hubbled

his position, trajectory, and age.

Their data were unambiguous.

"This man is a sage."

Commentary

There was no such thing as an "acoustical astronomer" when I was writing *Chronosphere* in 1997. The phrase was a playful way of uniting the image of our Falling Man — who is now singing his way to (or perhaps *through*) his tiny and individual end — with astronomy, the study of the very big picture. I wanted to introduce the idea that when we die, something enormous is happening, something with the potential to be as large as the universe itself.

To my delight, the practice of acoustical astronomy entered into to the world just a few years after I penned those words (though it has never been called that formally). I am referring to the study of gravitational waves — ripples in the fabric of spacetime, sometimes referred to as the "sound" of the universe. These were postulated but could never be observed until the completion of the LIGO project, two enormously sensitive astronomical detectors. When very big things happen somewhere in the cosmos, such as two black holes colliding, LIGO can hear the resulting boom.

LIGO tells us that our idea about space and time being different aspects of *one thing*, something visible to us mathematically but impossible to experience with our senses, must be right. Because we can now measure the sounds that travel through this spacetime like waves in the air or ripples on the water. And we can hear them, lapping gently against the shore of our planet.

Imagine this: the Chronosphere, conceived as the sphere of all human time around our planet, is a mere bubble in the vast, vast sea of spacetime. To fall out of the Chronosphere is to fall into the universal ocean. As the Singing Falling Man is about to find out.

56.

The Flying Woman hit a pocket of unexpected relativity. In theory, she began to fall. (In practice, not at all.)

Commentary

Until now, the mysterious Flying Woman has been both invincible and elusive. But as we approach the borders of what we know, or even what is knowable, the masterful edifice of physics — theories, measurements, certainty as well as our knowledge of fundamental uncertainty — begins to quiver ever so slightly. And this brings the Flying Woman into some kind of concert with the Falling Man. It seems they may share a destiny after all.

For physics has not tamed mystery. If anything, it has only deepened it. Consider the following experimental finding, published just seventeen days before I wrote these words. That same LIGO detector to which I referred in the previous commentary has now detected something quite different from what it was originally designed for. It has measured the essential bubbliness of our universe, the froth or foam roiling all around us, all the time.

In brief, particles wink in and out of existence all around us, all the time, at an unimaginably small-yet-vast scale. That is to say, the phenomenon is extremely tiny, but happens in enormous quantities. Really, our words for big and small have been thoroughly exhausted here. We are talking about scales so far below and above what humans can experience that only poetic approximations can help. Imagine, say, bubble bath bubbles, appearing and disappearing all around you, a hundred gazillion times per second. Now imagine that these bubbles are a gazillion

zillion times smaller than actual bubbles. That is the kind of universe we live in. And those bubbles are physically jiggling you, all the time.

Of course, you cannot see it, or feel it. But LIGO can sense such unimaginably small movements. It has measured these "quantum fluctuations" — the name scientists have given to the infinitesimally tiny frothiness all around us — as it jiggled something just as big as we are.

As physics progresses, theories do fall (though theory accurately predicted the presence of quantum fluctuations). But the universe itself does not. It just keeps doing what it has always done. And the mystery gets deeper.

57.

A black hole attracted the Singing Falling

Man. He whirled past the Flying Woman,

clasped her hand. They danced a jig

around the horizon, black-tunneled down,

white-light-fountained up into

a universe that wasn't quite so big.

Commentary

Whoosh! It happens — whatever "it" is. The cosmic fall is over. The Falling Man and Flying Woman are united at the exact time of his death. Or transformation. Or transportation into another realm.

Black holes have always fascinated me: deep wells in space whose gravitational pull is so strong that nothing, not even light, can escape. A fall into one of these is a one-way ticket to nowhere — at least, no "where" that we can understand. Doesn't that sound like death?

But perhaps death is not a trip to nowhere, but to somewhere, which is also the case with black holes. There is a possibility (so far unproven) that on the other end of a black hole is its polar opposite, a white hole, a kind of super-fountain of energy and light. Some have even postulated that white holes are how new universes form. Including, perhaps, our own.

Which is to say, the actions of this denouement are theoretically possible. Verse 57 might be more description than metaphor.

But the Falling Man's story is not yet complete.

58.

The Falling Man arrived

in time, or at the end of it.

It was neither night, nor day.

Not twilight; just some time

he couldn't say. He found

a book; his name

was on the cover.

He opened it.

His story

started over.

Commentary

Nearly two billion people in our world believe in reincarnation, the idea that after we die, we are reborn into a new life in this world. This belief is linked to a cyclical view of time: things repeat themselves, over and over, with slight variations. The business of being a soul involves improving oneself during this recurring process by making better choices, and thus creating better *karma* — a kind of cosmic accounting of one's holiness and purity. The ultimate goal is to become so pure that one is released from this cycle and united with the divine.

Over four billion people believe in a very different vision of the afterlife known as heaven and hell: you have one shot at life, and then at death, after being divinely judged, you may pass on to an eternity in paradise with the almighty — if you made the right choices concerning your beliefs and your behavior. If you did not, unending punishment and pain awaits you, in the company of the almighty's opposite number.

"Believe" might be too strong a word here, for the six billion people referred to above are the adherents of Buddhism, Hinduism, Christianity and Islam. And as we all know, it is perfectly possible to "belong" to one of these religions without actually believing in all of its tenets. So what does belief mean? How should we relate to our own beliefs?

Science tries to teach us that what we believe to be true about our universe makes no difference at all to the universe itself, which just does what it does, regardless of what we believe about it. Knowledge is built by observing what the universe is doing, testing theories, and constantly questioning our assumptions and beliefs. For history tells us that those assumptions and beliefs are often wrong.

But it is also true that what we believe about the nature of time, the purpose of life, and what happens to us when we die — questions for which science offers few satisfying answers — makes an enormous difference as to *how we live our lives*, what choices we make. It seems that the most important choice we make in life (if we are allowed a choice) is what to believe about life itself.

Verse 58 is not intended as an endorsement of any belief about life, death, or life after death. *Chronosphere* is just a story-poem, inspired by a man who faced death and had just a short time left to live. I chose to end the Falling Man's story "on time" — that is, before my friend died, but also before coming to the end of the 60-verse structure that I had set up for myself — with a kind of mélange of ideas from Buddhism and cosmology. It seems the Falling Man will be going through the cycle of life again, but in another, younger universe.

I do not exactly recall what my friend Wouter believed about life after death, if anything. He was a scientist and therefore skeptical about everything, including his own beliefs. But I felt that I could read these lines to him, on his deathbed, with a confidence that he would recognize the ideas and feel comfortable with them.

I also ended this verse with a reference to literature: for is it not so that books are a way of sending our consciousness into the future? Granted, it is but a small shard of our full awareness that can be coded into words, and the communication is also limited to sending, not receiving. But now you, the reader, know something of what I thought about my friend Wouter, and some of my reflections about how humanity views time, life and death, though I have been cagey about my personal beliefs.

I have also been keen to communicate that this commentary on *Chronosphere* is not a declaration of its meaning. The poem is what it is. Read into it what you want. As for the commentary, you can take it, or leave it.

But please, not yet.

59.

The Flying Woman watched

her hands spin round.

When both stretched overhead,

she made twelve ringing sounds.

Commentary

It was clear from the outset that the Flying Woman's story was an endless one. But is time endless? On that point, opinions sharply differ. The major religions all posit an end to time, or at least a great punctuation mark in its greatest cycles. An internet search on the phrase "Is time endless?" (the closest we can come to asking the Ultimate Authority) produces a babble of possible answers, ranging from hard-core theoretical physics to vague new-age philosophy. The best summary that one can provide on the state of human knowledge regarding this topic, at this time, is "maybe yes, maybe no."

But for a humble human pondering the cosmos, there is something comforting in knowing that for us, the question is moot. We have a certain allotment of this stuff we call time. Clocks, calendars, sundials, hourglasses, all the ways we have historically used to keep track of time tell us the same thing: it is not endless, not for any living creature on planet Earth. When it is all used up, it's gone. *We're* gone.

Time's up.

60.

Time is busy,

Time is lazy,

Time is just

a little crazy.

Commentary

I think I may have been lying to you. This was not my intention. Memory plays tricks on us. Let me try to set things right.

Writing *Chronosphere* kept me busy during a few weeks in 1997. In October of that year — before completing the first draft, but after writing some 45 verses — I penned a little prologue that did not survive the publication process. In fact, I had quite forgotten it, until it reappeared while drafting this commentary:

> *My hard drive is crashing, my memory's flea-bitten,*
> *I cannot remember a thing — so I've written*
> *these sixty-one verses and doggerel rhymes*
> *to commemorate the Balaton Meeting on Time*

I discovered this when plumbing my current hard drive for older versions of the poem, because I felt suddenly unsure of the timing of the project. When exactly had I written what? Perhaps the creation dates on old archive files would help — and they did. They established that I had conceived of the poem's 61-verse structure from the outset; that the Falling Man and Flying Woman were already fully formed images; and that I have been subtly (but unconsciously) bending the facts in what I have reported to you.

Most of Chronosphere was written just after the specific meeting of the Balaton Group on the subject of time (in September 1997) that I mentioned at the beginning. That is still true. But the first 45 verses seem to have been written *before* I realized that I wanted to dedicate the poem to Wouter.

Therefore I have no idea whether Wouter was truly the original inspiration for the image of the Falling Man. That link may have been something that occurred to me later, when I took up the draft again in March 1998, starting with verse 46 about "racing against time" — for that was the moment when it became clear that Wouter would not last long. I needed to finish *Chronosphere*, and present it to him, before it was too late.

It might be that a plunge into my journal — I have been writing a journal of my thoughts, feelings, experiences, dreams and reflections for over 40 years — would clear up this whole matter of timing and sequencing decisively. But what would be the point? I came to understand the image of the Falling Man, one of those poetic ideas that arrives in the mind after one opens the door and waits, thanks to Wouter and his struggle to face death with dignity, strength, vulnerability and integrity. The inspiration was at least retroactive, perhaps even unconscious from the start.

Perhaps I am just too lazy to clarify further details on timing, but frankly, I think this is enough. For 60 verses, you have been listening to me describe my friend, and my poem. It is time now that I allow Wouter to speak for himself. Here is an excerpt from a letter he wrote to our mutual friend Dana Meadows in December 1997, in which he attempted to convey everything he had learned from his intensive inner reflections while battling a terminal illness. Wouter related to Dana the "puzzle" of his own life in a marvelously clear and pithy way, recounting his key transitions and aha moments, and then summed it all up in a few sagacious sentences. I believe you will hear, as I do now, echoes of the Falling Man in these lines:

> *That's life, I guess — starting with talents, education and associated scars, stretching where you can, hiding where your fears are highest, exploiting events for the better, balancing outer and inner traveling, releasing more and more of the inessentials, enjoying more and more of the essentials of life.*

Can you imagine that I want to live longer to experience these insights more and more? But there is no way of knowing how long we have individually — and there is no "right to time." I know all that, but I still find it hard to swallow. And it makes me angry from time to time.

Reading this text for the first time in over two decades sparked another memory. For many months during Wouter's illness, a small group of us practiced something called "Wouter Time". At a particular time of day (I forget exactly when), we would think about Wouter and send him love. We did this in a synchronized way over numerous time zones, from the US to India. We were aiming to ease his suffering and to keep him going as long as possible. It does not matter whether one believes that such thoughts have a direct and material effect on the world. The important thing was that Wouter *knew* that we were doing this, and when. We thought about him, and he thought about us thinking about him, at the same time.

And that had an impact. For example, on the 10th of March, 1998, Wouter wrote again to Dana after a very up-and-down week. He was happy that the book he was writing about his vision of a sustainable world was about to be completed — just in time, considering his downward trajectory (Wouter's book is another association linked to the concluding image of a book in verse 58). But the tumors were spreading, and the pain was getting difficult to bear. He noted:

Today I felt very cold at Wouter time. But afterwards I was warm again, thanks to the warmth and love and energies sent by the Wouter time participants. Great!

Great, indeed. Though his Wikipedia entry is currently very short, I consider Wouter Biesiot to be a great man. He was certainly a great friend.

In the research that I did while composing this final commentary (which turned up Wouter's letter above), I also happened upon a surprise: a copy of an email sent from me to Dana Meadows, and other members of the "Wouter Time" group, following my last visit with him in April 1998. The day after he had finished his own book, the "pain and feeling of decline began in earnest," and he was being treated with morphine. "He faded out a time or two," I wrote, "but mostly it was still Wouter there, even more translucent, and more prone to the emotional or the witty." I had a few hours with him. I remember reading *Chronosphere* to him, and presenting him with a special handwritten copy in a handmade book — but I neglected to mention that in my letter:

> *We mostly just sat together. I felt it impossible not to touch him, to hold his arm or his shoulder or hand, rub them. He told me how he's been feeling. He gave himself about two weeks; he doesn't think he can stand the pain much longer than that.*
>
> *I told him, again, what an extraordinary contribution he's made to all of us. Not only intellectually, but emotionally and spiritually. No one should suffer what he's been suffering. But in doing it with such grace and realness and integrity and openness he has taught us all so much.*
>
> *We remembered our times together. I learned more about other important relationships in his life that had involved conflict and difficulty, all beautifully resolved now. We laughed a lot. He cried just once. I was teary on and off, but strangely joyful....*
>
> *When I left, he rose, and held me, and gave me one very light and tender kiss on the lips. It's a moment I will remember with intense vividness for the rest of my life.*

In fact, I had totally and completely forgotten about that moment.

Until just now.

61.

Lake Balaton

is a sheet of blue glass,

a window between after

and before. I dive

through the pane. On shore,

I hear the sound of thunder.

What's lost is clear; what's found

remains opaque. Down

at the bottom of the lake,

I see graffiti in the sand:

"The Noösphere

will understand."

Commentary

No further comment.

Coda

You can explore a web version of *The Chronosphere Commentary* at this address:

http://chronospherepoem.wordpress.com

The web version of the Commentary includes pauses, surprises, and links to other websites.

You can contact the author, Alan AtKisson, at this address:

chronosphere@atkisson.com

www.ingramcontent.com/pod-product-compliance
Lightning Source LLC
Chambersburg PA
CBHW070853050426
42453CB00012B/2168